From Fossil Fuels to Freedom

Embracing Renewable Energy for a Sustainable Future

Lois Powers

From Fossil Fuels to Freedom

TABLE OF CONTENTS

Chapter 1: The Transition to Renewable Energy

Defining Renewable Energy

Renewable energy, a term that has gained significant traction in recent years, refers to energy derived from natural sources that are replenished at a faster rate than they are consumed. Unlike fossil fuels, which are finite and contribute to environmental degradation, renewable energy sources offer a sustainable alternative that can meet the world's growing energy demands without depleting resources or causing harm to the planet. The essence of renewable energy lies in its ability to harness the power of nature—sunlight, wind, water, and organic matter—to generate electricity, heat, and fuel.

At the heart of renewable energy is the sun, the ultimate source of most renewable energy forms. Solar power, for instance, captures sunlight using photovoltaic cells or solar thermal systems to produce electricity or heat. The sun's energy also drives wind patterns, which can be harnessed by wind turbines to generate electricity. Similarly, the sun's heat causes water to evaporate, leading to precipitation that fills rivers and reservoirs, enabling hydropower generation. Even biomass energy, derived from organic materials, is a form of stored solar energy, as plants convert sunlight into chemical energy through photosynthesis.

The transition to renewable energy is driven by several key factors. Environmental concerns, particularly the need to reduce greenhouse gas emissions and combat climate change, are at the forefront. Fossil fuel combustion is a major contributor to carbon dioxide emissions, which trap heat in the atmosphere and lead to global warming. By shifting to renewable energy sources, we can significantly reduce our carbon footprint and mitigate the impacts of climate change. Additionally, renewable energy offers the promise of energy security and independence. Unlike fossil fuels, which are often concentrated in specific regions and subject to geopolitical tensions, renewable resources are abundant and widely distributed, reducing reliance on imports and enhancing energy resilience.

Economic considerations also play a crucial role in the transition to renewable energy. The cost of renewable technologies has been declining steadily, making them increasingly competitive with traditional energy sources. Advances in technology, economies of scale, and supportive policies have contributed to this trend, enabling wider adoption and investment in renewable energy projects. Moreover, the renewable energy sector has become a significant driver of job creation, offering employment opportunities in manufacturing, installation, maintenance, and research and development.

Despite the numerous advantages, the transition to renewable energy is not without challenges. One of the primary obstacles is the intermittent nature of some renewable sources, such as solar and wind power, which depend on weather conditions and time of day. This variability necessitates the development of energy storage solutions and grid management strategies to ensure a reliable and stable energy supply. Battery technology, pumped hydro storage, and smart grid systems are among the innovations being explored to address these challenges.

Another challenge is the need for substantial infrastructure investment. Transitioning to renewable energy requires upgrading existing energy systems, building new facilities, and expanding transmission networks to accommodate decentralized energy generation. This requires significant financial resources and coordinated efforts from governments, private sector stakeholders, and communities. Additionally, regulatory frameworks and market structures must be adapted to support the integration of renewable energy and incentivize investment.

Public perception and acceptance are also critical to the success of renewable energy projects. Community engagement and education are essential to address concerns and misconceptions about renewable technologies, such as their environmental impact, aesthetic considerations, and potential effects on local economies. By fostering dialogue and collaboration, stakeholders can build trust and support for renewable energy initiatives.

Innovation and technology play a pivotal role in overcoming these challenges and unlocking the full potential of renewable energy. Research and development efforts are focused on improving the efficiency and cost-effectiveness of renewable technologies, as well as exploring new and emerging energy sources. For example, advancements in solar panel materials, wind turbine design, and bioenergy conversion processes are enhancing the performance and scalability of these technologies. Additionally, digital technologies, such as artificial intelligence and the Internet of Things, are being leveraged to optimize energy systems, enhance grid management, and enable smart energy solutions.

Successful transitions to renewable energy can be seen in various case studies around the world. Countries like Denmark and Germany have made significant strides in integrating renewable energy into their national energy mix, achieving high shares of renewable electricity generation. These successes are attributed to strong policy frameworks, public-private partnerships, and a commitment to innovation and sustainability. Similarly, cities and regions are implementing ambitious renewable energy targets and initiatives, demonstrating the feasibility and benefits of transitioning to a low-carbon future.

The journey towards renewable energy is a complex and multifaceted endeavor, requiring collaboration, innovation, and determination. By embracing renewable energy, we can create a more sustainable, resilient, and equitable energy system that meets the needs of present and future generations. As we continue to explore and harness the power of nature, we have the opportunity to redefine our relationship with energy and pave the way for a cleaner, greener world.

Key Drivers of the Transition

The shift towards renewable energy is propelled by a confluence of factors that are reshaping the global energy landscape. At the forefront of this transition is the urgent need to address climate change, a challenge that has galvanized governments, businesses, and individuals worldwide. The burning of fossil fuels is a primary contributor to greenhouse gas emissions, which trap heat in the atmosphere and lead to global warming. As the impacts of climate change become increasingly evident—rising sea levels, extreme weather events, and biodiversity loss—the imperative to reduce carbon emissions has never been more pressing. Renewable energy offers a viable solution, providing a pathway to decarbonize the energy sector and mitigate the adverse effects of climate change.

Economic considerations are equally influential in driving the transition to renewable energy. The cost of renewable technologies has plummeted over the past decade, making them more competitive with traditional energy sources. Solar and wind power, in particular, have seen dramatic reductions in cost due to technological advancements, economies of scale, and increased market competition. As a result, renewable energy projects are becoming more financially attractive, drawing investment from both public and private sectors. This economic viability is further bolstered by the potential for job creation in the renewable energy industry, which offers employment opportunities in manufacturing, installation, maintenance, and research and development.

Energy security is another critical driver of the transition to renewable energy. Fossil fuels are often concentrated in specific regions, leading to geopolitical tensions and supply vulnerabilities. In contrast, renewable resources are abundant and widely distributed, reducing reliance on imports and enhancing energy independence. By diversifying energy sources and decentralizing energy production, countries can strengthen their energy security and resilience against external shocks. This is particularly important in the context of fluctuating fossil fuel prices and the geopolitical instability that can disrupt energy supplies.

Technological innovation plays a pivotal role in facilitating the transition to renewable energy. Advances in technology are improving the efficiency, reliability, and scalability of renewable energy systems. For example, improvements in solar panel materials and manufacturing processes have increased the efficiency of photovoltaic cells, enabling them to capture more sunlight and generate more electricity. Similarly, innovations in wind turbine design have enhanced their performance, allowing them to operate in a wider range of wind conditions and generate more power. Energy storage technologies, such as batteries and pumped hydro storage, are also evolving, providing solutions to the intermittency challenges associated with some renewable energy sources.

Policy and regulatory frameworks are essential in driving the transition to renewable energy. Governments play a crucial role in creating an enabling environment for renewable energy development through supportive policies, incentives, and regulations. Feed-in tariffs, tax credits, and renewable portfolio standards are among the mechanisms used to encourage investment in renewable energy projects. Additionally, international agreements, such as the Paris Agreement, have set ambitious targets for reducing carbon emissions and increasing the share of renewable energy in the global energy mix. These policy measures signal a commitment to sustainability and provide a clear direction for the energy transition.

Public awareness and consumer demand are also significant drivers of the transition to renewable energy. As awareness of environmental issues grows, consumers are increasingly seeking sustainable and eco-friendly products and services. This shift in consumer preferences is influencing businesses to adopt renewable energy solutions and reduce their carbon footprint. Companies are recognizing the value of sustainability as a competitive advantage and are integrating renewable energy into their operations and supply chains. Furthermore, community-led initiatives and grassroots movements are advocating for renewable energy adoption, demonstrating the power of collective action in driving change.

The transition to renewable energy is not without its challenges, but the drivers propelling this shift are strong and multifaceted. By addressing climate change, enhancing energy security, and leveraging economic opportunities, renewable energy offers a compelling vision for a sustainable future. Technological innovation, supportive policies, and public engagement are key to overcoming barriers and accelerating the transition. As we navigate this transformative journey, the collective efforts of governments, businesses, and individuals will be crucial in realizing the full potential of renewable energy and creating a cleaner, greener world for generations to come.

Challenges and Opportunities

Navigating the transition from fossil fuels to renewable energy presents a complex landscape filled with both formidable challenges and promising opportunities. The journey is akin to steering a ship through turbulent waters, where the destination is clear, but the path is fraught with obstacles that require careful navigation and strategic foresight.

One of the most significant challenges is the intermittency of renewable energy sources such as solar and wind. Unlike fossil fuels, which can provide a constant supply of energy, solar and wind power are dependent on weather conditions and time of day. This variability poses a challenge for maintaining a stable and reliable energy supply. To address this, advancements in energy storage technologies are crucial. Batteries, pumped hydro storage, and other innovative solutions are being developed to store excess energy generated during peak production times and release it when demand is high or production is low. These technologies are essential for smoothing out the fluctuations in energy supply and ensuring a consistent flow of electricity to the grid.

Another challenge lies in the existing energy infrastructure, which has been built around centralized fossil fuel power plants. Transitioning to renewable energy requires significant investment in new infrastructure, including decentralized energy generation systems, upgraded transmission networks, and smart grid technologies. This transformation demands substantial financial resources and coordinated efforts from governments, utilities, and private sector stakeholders. The scale of investment needed can be daunting, but it also presents an opportunity for economic growth and job creation. The renewable energy sector has the potential to generate millions of jobs in manufacturing, installation, maintenance, and research and development, providing a boost to local economies and fostering innovation.

The regulatory and policy environment is another critical factor influencing the transition to renewable energy. In many regions, existing policies and market structures are designed to support fossil fuel industries, creating barriers to the adoption of renewable energy. Reforming these frameworks to incentivize renewable energy development and create a level playing field is essential. Governments can play a pivotal role by implementing supportive policies such as feed-in tariffs, tax credits, and renewable portfolio standards. These measures can encourage investment in renewable energy projects and accelerate the transition. Additionally, international cooperation and agreements, such as the Paris Agreement, provide a platform for countries to commit to reducing carbon emissions and increasing the share of renewable energy in their energy mix.

Public perception and acceptance are also crucial in the transition to renewable energy. While there is growing awareness of the environmental benefits of renewable energy, there are still misconceptions and resistance in some communities. Concerns about the visual impact of wind turbines, the land use requirements of solar farms, and the potential effects on local wildlife can hinder the development of renewable energy projects. Engaging with communities, providing transparent information, and involving local stakeholders in decision-making processes are essential strategies for building trust and gaining public support. Education and outreach efforts can help dispel myths and highlight the benefits of renewable energy, fostering a culture of sustainability and environmental stewardship.

Despite these challenges, the opportunities presented by the transition to renewable energy are immense. Renewable energy offers a pathway to a more sustainable and resilient energy system, reducing reliance on finite fossil fuels and decreasing greenhouse gas emissions. By harnessing the power of nature, we can create a cleaner, healthier environment for future generations. The transition also presents an opportunity to democratize energy production, empowering individuals and communities to generate their own energy and participate in the energy market. This shift towards decentralized energy systems can enhance energy security, reduce energy costs, and increase access to electricity in remote and underserved areas.

Innovation and technology are key drivers of the opportunities in the renewable energy sector. Advances in solar panel efficiency, wind turbine design, and bioenergy conversion processes are enhancing the performance and scalability of renewable technologies. Digital technologies, such as smart grids and energy management systems, are enabling more efficient and flexible energy systems, optimizing the integration of renewable energy into the grid. These innovations are not only improving the viability of renewable energy but also opening up new possibilities for energy solutions that were previously unimaginable.

The transition to renewable energy is a journey filled with both challenges and opportunities. By addressing the obstacles and seizing the opportunities, we can create a more sustainable, equitable, and prosperous energy future. The collective efforts of governments, businesses, communities, and individuals will be crucial in navigating this transition and realizing the full potential of renewable energy. As we move forward, the vision of a world powered by clean, renewable energy is within reach, offering hope and promise for a brighter tomorrow.

The Role of Innovation and Technology

Innovation and technology are the twin engines propelling the transition from fossil fuels to renewable energy. As the world grapples with the pressing need to reduce carbon emissions and combat climate change, technological advancements are paving the way for a cleaner, more sustainable energy future. The role of innovation in this transition cannot be overstated, as it is the key to overcoming the challenges associated with renewable energy and unlocking its full potential.

One of the most significant areas of innovation is in the field of solar energy. Photovoltaic (PV) technology, which converts sunlight directly into electricity, has seen remarkable advancements in recent years. Researchers are continually developing new materials and manufacturing techniques to increase the efficiency of solar panels. For instance, the use of perovskite materials has shown promise in achieving higher efficiency rates than traditional silicon-based panels. These materials are not only more efficient but also cheaper to produce, making solar energy more accessible and affordable. Additionally, innovations in solar panel design, such as bifacial panels that capture sunlight from both sides, are further enhancing energy capture and efficiency.

Wind energy has also benefited from technological advancements. Modern wind turbines are a testament to engineering ingenuity, with taller towers and longer blades that capture more wind and generate more power. Innovations in turbine design have led to increased capacity and efficiency, allowing wind farms to produce more electricity with fewer turbines. Offshore wind technology is another area of rapid development, with floating wind turbines opening up new possibilities for harnessing wind energy in deeper waters. These advancements are expanding the potential for wind energy, making it a more viable option for a wider range of locations.

Energy storage is a critical component of the renewable energy landscape, addressing the intermittency challenges associated with solar and wind power. Technological innovations in battery storage are revolutionizing the way we store and use energy. Lithium-ion batteries, commonly used in electric vehicles and portable electronics, are being adapted for grid-scale energy storage, providing a reliable solution for balancing supply and demand. Researchers are also exploring alternative battery technologies, such as solid-state batteries and flow batteries, which offer the potential for higher energy density and longer lifespans. These advancements in energy storage are crucial for ensuring a stable and reliable energy supply, even when the sun isn't shining or the wind isn't blowing.

The integration of digital technologies is another area where innovation is playing a transformative role. Smart grids, which use digital communication technology to monitor and manage energy flows, are enabling more efficient and flexible energy systems. These grids can dynamically adjust to changes in energy supply and demand, optimizing the use of renewable energy and reducing waste. The Internet of Things (IoT) is also being leveraged to create smart homes and buildings that can manage their energy consumption more effectively. By connecting appliances, lighting, and heating systems to the grid, these smart systems can automatically adjust energy use based on real-time data, reducing energy consumption and costs.

Bioenergy, derived from organic materials such as plants and waste, is another area where innovation is driving progress. Advances in bioenergy conversion technologies are improving the efficiency and sustainability of biofuel production. For example, the development of advanced biofuels, such as cellulosic ethanol and algae-based fuels, is providing cleaner alternatives to traditional fossil fuels. These biofuels can be produced from non-food crops and waste materials, reducing competition with food production and minimizing environmental impact. Innovations in biogas production, which captures methane from organic waste, are also contributing to a more sustainable energy system by turning waste into a valuable energy resource.

Hydropower, one of the oldest forms of renewable energy, is also benefiting from technological advancements. Innovations in turbine design and materials are increasing the efficiency and environmental sustainability of hydropower systems. Small-scale and micro-hydropower technologies are expanding the potential for hydropower in remote and off-grid locations, providing clean energy to communities that may not have access to traditional energy sources. Additionally, the development of pumped hydro storage, which uses excess electricity to pump water uphill for later use, is providing a valuable solution for energy storage and grid stability.

The role of innovation and technology in the transition to renewable energy is multifaceted and dynamic. By pushing the boundaries of what is possible, these advancements are not only making renewable energy more efficient and cost-effective but also opening up new possibilities for energy solutions that were previously unimaginable. As we continue to innovate and explore new technologies, the potential for renewable energy to transform our energy systems and create a more sustainable future is limitless. The journey is ongoing, and the opportunities are vast, as we harness the power of human ingenuity to build a cleaner, greener world.

Case Studies: Successful Transitions

Across the globe, numerous countries and regions have embarked on the journey towards renewable energy, each with its unique approach and set of challenges. These case studies of successful transitions offer valuable insights into the strategies and practices that have enabled them to achieve significant progress in integrating renewable energy into their energy systems.

Denmark stands as a beacon of success in the renewable energy transition. With a commitment to sustainability and energy independence, Denmark has become a global leader in wind energy. The country's journey began in the 1970s, driven by the oil crisis and a desire to reduce reliance on fossil fuels. Denmark invested heavily in wind power technology, fostering innovation and creating a robust domestic industry. Today, wind energy accounts for nearly half of Denmark's electricity consumption, a testament to the country's strategic planning and policy support. The Danish government implemented favorable policies, such as feed-in tariffs and subsidies, to encourage investment in wind energy. Additionally, Denmark's focus on community ownership and involvement has played a crucial role in gaining public support and acceptance for wind projects. By empowering local communities to invest in and benefit from wind farms, Denmark has created a model of inclusive and sustainable energy development.

Germany's Energiewende, or "energy transition," is another remarkable example of a successful shift towards renewable energy. Germany's ambitious plan aims to phase out nuclear power and significantly reduce carbon emissions by increasing the share of renewables in its energy mix. The country has made substantial investments in solar and wind energy, supported by a comprehensive policy framework that includes feed-in tariffs, renewable energy targets, and grid expansion initiatives. Germany's decentralized energy system, characterized by a high level of citizen participation and ownership, has been instrumental in driving the transition. The proliferation of small-scale solar installations on rooftops and community-owned wind projects has democratized energy production and fostered a culture of sustainability. Despite challenges such as grid integration and fluctuating energy prices, Germany's commitment to renewable energy remains steadfast, serving as an inspiration for other nations.

Costa Rica offers a compelling case study of a country that has achieved near-total reliance on renewable energy. Blessed with abundant natural resources, Costa Rica has harnessed its rivers, wind, and geothermal potential to generate clean electricity. The country's commitment to environmental conservation and sustainable development has been a driving force behind its renewable energy success. Costa Rica's government has prioritized investments in renewable infrastructure, supported by policies that promote energy efficiency and conservation. The country's integrated approach, which combines hydropower, wind, solar, and geothermal energy, has enabled it to achieve a renewable energy share of over 99% in its electricity generation. Costa Rica's success demonstrates the power of political will and strategic planning in achieving ambitious renewable energy goals.

In the United States, the state of California has emerged as a leader in renewable energy adoption. California's progressive energy policies and commitment to reducing greenhouse gas emissions have propelled the state to the forefront of the renewable energy movement. The state's Renewable Portfolio Standard mandates that a significant portion of electricity must come from renewable sources, driving investment in solar, wind, and other clean energy technologies. California's innovative approach to energy storage and grid management has also been a key factor in its success. The state's investment in battery storage and demand response programs has enhanced grid reliability and facilitated the integration of intermittent renewable energy sources. California's experience highlights the importance of policy leadership and technological innovation in advancing the renewable energy transition.

The island nation of Iceland provides a unique example of a successful transition to renewable energy, driven by its abundant geothermal resources. Iceland's geothermal energy journey began in the early 20th century, with the development of geothermal heating systems for homes and businesses. Today, geothermal energy accounts for a significant portion of Iceland's electricity and heating needs, complemented by hydropower. Iceland's commitment to renewable energy is rooted in its desire for energy independence and environmental sustainability. The country's investment in geothermal research and development has positioned it as a global leader in geothermal technology, attracting international collaboration and knowledge sharing. Iceland's experience underscores the potential of leveraging local resources and expertise to achieve a sustainable energy future.

These case studies illustrate that successful transitions to renewable energy are characterized by a combination of strategic planning, supportive policies, technological innovation, and community engagement. Each example offers valuable lessons for other regions and countries seeking to embark on their renewable energy journey. By learning from these successes, we can accelerate the global transition to a cleaner,

more sustainable energy system, paving the way for a brighter
and more resilient future.

How Solar Energy Works

Solar energy, a cornerstone of the renewable energy revolution, harnesses the power of the sun to generate electricity and heat. Understanding how solar energy works involves delving into the science and technology that capture and convert sunlight into usable energy. At its core, solar energy relies on the principles of photovoltaic (PV) technology and solar thermal systems, each offering unique methods for tapping into the sun's abundant energy.

Photovoltaic technology is the most common method for converting sunlight into electricity. The process begins with solar panels, which are composed of numerous solar cells made from semiconductor materials, typically silicon. These cells are designed to absorb photons, the particles of light that carry energy from the sun. When photons strike the surface of a solar cell, they transfer their energy to electrons within the semiconductor material, knocking them loose from their atoms. This movement of electrons generates an electric current, which can be captured and directed for use in homes, businesses, and the grid.

The efficiency of photovoltaic cells is a critical factor in determining how much electricity can be generated from a given amount of sunlight. Advances in materials science and engineering have led to the development of high-efficiency solar cells, such as those made from monocrystalline silicon, polycrystalline silicon, and thin-film technologies. Each type of solar cell has its advantages and trade-offs, with monocrystalline cells offering higher efficiency and durability, while thin-film cells provide flexibility and lower production costs.

Solar panels are typically installed in arrays, which are groups of panels connected together to increase the overall power output. These arrays can be mounted on rooftops, integrated into building materials, or deployed in large-scale solar farms. The orientation and angle of the panels are crucial for maximizing energy capture, as they need to be positioned to receive the most direct sunlight throughout the day. In some cases, solar tracking systems are used to adjust the position of the panels, following the sun's path across the sky to optimize energy generation.

Once the solar panels generate electricity, it is typically in the form of direct current (DC). However, most homes and businesses use alternating current (AC) for their electrical systems. To convert the DC electricity into AC, an inverter is used. Inverters are essential components of solar energy systems, ensuring that the electricity produced by the solar panels is compatible with the electrical grid and household appliances. Modern inverters also come equipped with smart technology, allowing for real-time monitoring and optimization of energy production.

In addition to photovoltaic systems, solar thermal technology offers another method for harnessing solar energy. Solar thermal systems capture and concentrate sunlight to produce heat, which can be used for various applications, including water heating, space heating, and even electricity generation. One common type of solar thermal system is the solar water heater, which uses solar collectors to absorb sunlight and transfer the heat to water stored in a tank. This heated water can then be used for domestic purposes, reducing the need for conventional water heating methods.

For larger-scale applications, concentrated solar power (CSP) systems are employed. CSP systems use mirrors or lenses to focus sunlight onto a small area, creating intense heat that can be used to produce steam. This steam drives a turbine connected to a generator, producing electricity in a manner similar to traditional power plants. CSP systems are particularly effective in regions with high solar insolation, where they can provide a reliable and consistent source of renewable energy.

Energy storage is a critical component of solar energy systems, addressing the challenge of intermittency when the sun is not shining. Batteries are commonly used to store excess electricity generated during peak sunlight hours, allowing it to be used during periods of low solar production, such as at night or on cloudy days. Advances in battery technology, such as lithium-ion and flow batteries, are enhancing the capacity and efficiency of energy storage solutions, making solar energy more reliable and accessible.

The integration of solar energy into the grid presents both opportunities and challenges. On one hand, solar energy contributes to a cleaner and more sustainable energy mix, reducing reliance on fossil fuels and decreasing greenhouse gas emissions. On the other hand, the variability of solar energy production requires careful management to ensure grid stability and reliability. Smart grid technologies and demand response programs are being developed to address these challenges, enabling more efficient and flexible energy systems.

Solar energy offers a multitude of benefits, from reducing energy costs and carbon emissions to increasing energy independence and security. By understanding how solar energy works and the technologies that enable its capture and conversion, individuals and communities can make informed decisions about adopting solar solutions. As the world continues to embrace renewable energy, solar power stands out as a vital component of a sustainable energy future, harnessing the limitless potential of the sun to power our lives and protect our planet.

Advances in Solar Technology

Solar technology has undergone a remarkable evolution, transforming from a niche energy source into a cornerstone of the global renewable energy landscape. This transformation is driven by continuous advances in solar technology, which have significantly improved the efficiency, affordability, and versatility of solar energy systems. These innovations are not only making solar power more accessible but also expanding its potential applications across various sectors.

One of the most significant advancements in solar technology is the development of high-efficiency photovoltaic (PV) cells. Traditional silicon-based solar cells have been the industry standard for decades, but recent breakthroughs have led to the creation of new materials and cell designs that offer higher efficiency rates. Perovskite solar cells, for example, have emerged as a promising alternative to silicon cells. These cells are made from a class of materials known as perovskites, which have a unique crystal structure that allows for efficient light absorption and charge transport. Perovskite solar cells have achieved impressive efficiency gains in laboratory settings, and ongoing research aims to improve their stability and scalability for commercial use.

Another exciting development in solar technology is the advent of bifacial solar panels. Unlike traditional solar panels that capture sunlight on one side, bifacial panels are designed to absorb light from both the front and back surfaces. This dual-sided design allows them to capture reflected sunlight from the ground or surrounding surfaces, increasing their overall energy output. Bifacial panels are particularly effective in environments with high albedo, such as snowy or sandy areas, where the ground reflects a significant amount of sunlight. By harnessing additional light, bifacial panels can achieve higher energy yields without requiring additional space, making them an attractive option for both residential and utility-scale installations.

Thin-film solar technology represents another area of significant advancement. Thin-film solar cells are made by depositing one or more layers of photovoltaic material onto a substrate, such as glass, plastic, or metal. These cells are lightweight, flexible, and can be manufactured at a lower cost than traditional silicon cells. Thin-film technology is particularly well-suited for applications where flexibility and weight are critical factors, such as in building-integrated photovoltaics (BIPV) and portable solar devices. BIPV systems integrate solar cells directly into building materials, such as windows, facades, and roofing, allowing for seamless incorporation of solar energy generation into the built environment. This approach not only reduces the need for separate solar installations but also enhances the aesthetic appeal of solar technology.

Concentrated solar power (CSP) technology has also seen significant advancements, offering a complementary approach to photovoltaic systems. CSP systems use mirrors or lenses to concentrate sunlight onto a small area, generating heat that can be used to produce electricity. Recent innovations in CSP technology include the development of advanced heat transfer fluids and thermal energy storage systems, which enhance the efficiency and reliability of CSP plants. These advancements enable CSP systems to provide a stable and dispatchable source of renewable energy, even when the sun is not shining. By storing thermal energy in materials such as molten salts, CSP plants can continue to generate electricity after sunset, providing a valuable solution for grid stability and energy security.

The integration of digital technologies is revolutionizing the way solar energy systems are managed and optimized. Smart inverters, for example, are equipped with advanced communication capabilities that allow them to interact with the grid and other energy systems. These inverters can dynamically adjust the flow of electricity based on real-time data, optimizing energy production and consumption. Additionally, the use of artificial intelligence and machine learning algorithms is enabling more accurate forecasting of solar energy generation, improving grid integration and reducing the need for backup power sources.

Energy storage solutions are also playing a crucial role in advancing solar technology. The development of high-capacity batteries, such as lithium-ion and solid-state batteries, is enhancing the ability to store and utilize solar energy effectively. These storage systems allow excess solar energy generated during peak sunlight hours to be stored and used during periods of low solar production, such as at night or on cloudy days. By providing a reliable and flexible energy supply, energy storage solutions are overcoming one of the key challenges associated with solar energy and enabling greater penetration of solar power into the energy mix.

The rapid pace of innovation in solar technology is opening up new possibilities for its application across various sectors. In agriculture, for example, solar-powered irrigation systems are providing a sustainable solution for water management, reducing reliance on diesel pumps and lowering operational costs. In transportation, solar panels are being integrated into electric vehicles and charging stations, offering a clean and renewable source of energy for mobility. The potential for solar technology to transform industries and improve quality of life is vast, and continued advancements will only expand its impact.

As solar technology continues to evolve, it is clear that the future of energy is bright. The ongoing advancements in efficiency, cost reduction, and integration are making solar power an increasingly viable and attractive option for meeting the world's energy needs. By harnessing the power of the sun, we can create a more sustainable and resilient energy system, paving the way for a cleaner and greener future. The journey of solar technology is far from over, and the innovations yet to come hold the promise of even greater achievements in the quest for renewable energy.

Economic and Environmental Benefits

The transition to renewable energy, particularly solar power, offers a multitude of economic and environmental benefits that are reshaping the global energy landscape. As countries and communities increasingly embrace solar technology, the positive impacts on economies and ecosystems are becoming more evident, driving further investment and innovation in this sustainable energy source.

Economically, solar energy presents a compelling case for reducing energy costs and fostering economic growth. One of the most significant advantages is the decreasing cost of solar technology. Over the past decade, the cost of solar panels has plummeted, making solar energy more affordable and accessible than ever before. This reduction in cost is largely due to advancements in manufacturing processes, economies of scale, and increased competition in the solar industry. As a result, solar power has become a cost-competitive alternative to traditional fossil fuels, offering significant savings on electricity bills for both residential and commercial consumers.

The solar industry is also a powerful engine for job creation and economic development. As the demand for solar installations grows, so does the need for skilled workers in manufacturing, installation, maintenance, and research and development. The solar sector has become one of the fastest-growing job markets, providing employment opportunities across a wide range of skill levels and disciplines. This growth not only supports local economies but also contributes to a more resilient and diversified workforce. Moreover, the decentralized nature of solar energy allows for economic benefits to be distributed more evenly, empowering communities to generate their own energy and participate in the energy market.

On a larger scale, the adoption of solar energy can enhance energy security and reduce dependence on imported fossil fuels. By harnessing the power of the sun, countries can decrease their reliance on volatile energy markets and geopolitical tensions associated with fossil fuel imports. This shift towards energy independence not only strengthens national security but also stabilizes energy prices, providing a more predictable and sustainable energy future.

Environmentally, the benefits of solar energy are profound and far-reaching. One of the most significant advantages is the reduction of greenhouse gas emissions. Unlike fossil fuels, solar power generates electricity without releasing carbon dioxide or other harmful pollutants into the atmosphere. By replacing coal, oil, and natural gas with solar energy, we can significantly decrease the carbon footprint of our energy systems, mitigating the impacts of climate change and improving air quality. This reduction in emissions is crucial for meeting international climate targets and protecting the planet for future generations.

Solar energy also contributes to the conservation of natural resources. Traditional energy production often involves the extraction and consumption of finite resources, such as coal, oil, and natural gas. In contrast, solar power relies on the sun, an abundant and inexhaustible resource. By tapping into this renewable source of energy, we can reduce the strain on natural resources and preserve ecosystems for biodiversity and future use. Additionally, solar energy systems have a relatively low environmental impact compared to other forms of energy production, with minimal land and water use and no emissions during operation.

The integration of solar energy into the grid also supports the development of more resilient and adaptable energy systems. As climate change increases the frequency and severity of extreme weather events, the need for robust and flexible energy infrastructure becomes more critical. Solar power, with its decentralized nature and ability to be deployed in diverse locations, enhances the resilience of energy systems by reducing the risk of widespread outages and increasing the capacity for local energy generation. This adaptability is particularly valuable in remote or underserved areas, where access to reliable energy can be a challenge.

Furthermore, solar energy can play a vital role in promoting environmental justice and equity. Historically, marginalized communities have borne the brunt of pollution and environmental degradation associated with fossil fuel production and consumption. By transitioning to clean and renewable energy sources like solar power, we can reduce the environmental burdens on these communities and promote a more equitable distribution of energy resources. Community solar projects, for example, allow individuals and families who may not have the means to install solar panels on their own properties to benefit from shared solar installations, reducing energy costs and increasing access to clean energy.

The economic and environmental benefits of solar energy are undeniable, offering a pathway to a more sustainable and prosperous future. As we continue to innovate and invest in solar technology, the potential for positive change will only grow, transforming our energy systems and improving the quality of life for people around the world. By embracing solar power, we can create a cleaner, healthier, and more equitable world, harnessing the power of the sun to drive progress and protect our planet.

Overcoming Barriers to Adoption

The journey toward widespread adoption of renewable energy, particularly solar power, is fraught with challenges that must be addressed to unlock its full potential. Overcoming these barriers requires a multifaceted approach that involves technological innovation, policy support, financial mechanisms, and public engagement. By understanding and addressing these obstacles, we can accelerate the transition to a sustainable energy future.

One of the primary barriers to the adoption of solar energy is the initial cost of installation. While the cost of solar panels has decreased significantly over the years, the upfront investment required for purchasing and installing a solar energy system can still be prohibitive for many individuals and businesses. To address this challenge, innovative financing options have emerged, such as solar leasing, power purchase agreements (PPAs), and community solar programs. These models allow consumers to access solar energy without the burden of high initial costs, making it more affordable and accessible. By spreading the cost over time or sharing the benefits of a larger solar installation, these financial mechanisms can lower the entry barrier and encourage broader adoption.

Another significant challenge is the intermittency of solar energy, as it is dependent on sunlight availability. This variability can pose difficulties for grid integration and reliability, particularly in regions with less consistent sunlight. Advances in energy storage technology are crucial for overcoming this barrier. By storing excess solar energy generated during peak sunlight hours, batteries and other storage solutions can provide a steady supply of electricity even when the sun is not shining. Continued research and development in battery technology, such as lithium-ion and solid-state batteries, are enhancing storage capacity and efficiency, making solar energy a more reliable and viable option.

Grid infrastructure and integration also present challenges for the widespread adoption of solar energy. Many existing grids were designed for centralized power generation and may require upgrades to accommodate the decentralized nature of solar power. Smart grid technologies, which use digital communication to monitor and manage energy flows, are essential for optimizing the integration of solar energy into the grid. These technologies enable more efficient energy distribution, demand response, and real-time monitoring, ensuring grid stability and reliability. Investment in grid modernization and expansion is necessary to support the growing share of solar energy in the energy mix.

Policy and regulatory frameworks play a critical role in facilitating or hindering the adoption of solar energy. Supportive policies, such as feed-in tariffs, tax incentives, and renewable energy targets, can create a favorable environment for solar development. However, inconsistent or unclear regulations can create uncertainty and deter investment. Policymakers must establish clear and stable regulatory frameworks that encourage investment in solar energy and provide long-term certainty for developers and consumers. Collaboration between government, industry, and stakeholders is essential for creating policies that balance economic, environmental, and social considerations.

Public perception and awareness are also important factors in the adoption of solar energy. Misconceptions about the reliability, cost, and environmental impact of solar power can hinder its acceptance and uptake. Public education and outreach efforts are crucial for dispelling myths and raising awareness about the benefits of solar energy. By providing accurate information and highlighting successful case studies, we can build public support and encourage more individuals and communities to embrace solar solutions. Community engagement and participation in solar projects can also foster a sense of ownership and empowerment, driving further adoption.

Land use and environmental considerations can pose challenges for large-scale solar installations. Solar farms require significant land area, which can lead to conflicts with agricultural, conservation, or recreational land uses. Careful site selection and planning are essential to minimize environmental impact and balance competing land use interests. Innovative approaches, such as agrivoltaics, which combine solar energy production with agriculture, offer solutions for maximizing land use efficiency and creating synergies between energy and food production.

Technological innovation is a key driver in overcoming barriers to solar energy adoption. Continued research and development in solar technology are leading to more efficient, cost-effective, and versatile solutions. Advances in materials science, such as the development of perovskite solar cells and bifacial panels, are enhancing the efficiency and performance of solar systems. Additionally, the integration of digital technologies, such as artificial intelligence and machine learning, is optimizing energy management and grid integration. By fostering a culture of innovation and collaboration, we can continue to push the boundaries of what is possible with solar energy.

The transition to solar energy is a complex and multifaceted process that requires coordinated efforts across various sectors and stakeholders. By addressing the barriers to adoption through innovative solutions, supportive policies, and public engagement, we can accelerate the shift towards a sustainable energy future. The potential for solar energy to transform our energy systems and improve quality of life is immense, and by overcoming these challenges, we can unlock its full potential and create a cleaner, more resilient world.

Inspiring Solar Success Stories

Across the globe, solar energy has sparked a revolution, transforming communities and industries with its promise of clean, sustainable power. The stories of those who have embraced solar technology are as diverse as they are inspiring, showcasing the potential of solar energy to drive positive change and innovation. These success stories not only highlight the technical achievements of solar pioneers but also underscore the social and economic impacts of adopting renewable energy solutions.

In the heart of Africa, the small village of Kitonyoni in Kenya stands as a testament to the transformative power of solar energy. Before the introduction of solar power, Kitonyoni's residents relied on kerosene lamps for lighting, which were not only costly but also posed health risks due to indoor air pollution. The installation of a solar microgrid changed everything. Funded by a combination of local contributions and international support, the microgrid now provides reliable electricity to the entire village. This access to clean energy has revolutionized daily life, enabling businesses to extend their operating hours, improving educational opportunities with better lighting in schools, and enhancing healthcare services with powered medical equipment. The success of Kitonyoni's solar project has inspired neighboring communities to pursue similar initiatives, creating a ripple effect of sustainable development across the region.

In the bustling city of San Diego, California, a different kind of solar success story is unfolding. The city has emerged as a leader in urban solar adoption, driven by a combination of progressive policies, community engagement, and innovative technology. San Diego's commitment to renewable energy is evident in its ambitious Climate Action Plan, which aims to achieve 100% renewable electricity by 2035. The city's solar initiatives have been bolstered by supportive policies, such as streamlined permitting processes and financial incentives for solar installations. As a result, San Diego boasts one of the highest per capita rates of solar installations in the United States. The widespread adoption of solar energy has not only reduced the city's carbon footprint but also created thousands of jobs in the solar industry, contributing to economic growth and resilience.

In the remote Himalayan region of Ladakh, India, solar energy is providing a lifeline for communities living in one of the harshest environments on Earth. With limited access to the national grid, many villages in Ladakh have historically relied on diesel generators for electricity, which are both expensive and environmentally damaging. The introduction of solar power has been a game-changer for these communities. Solar microgrids and home solar systems now provide reliable and affordable electricity, reducing dependence on diesel and improving quality of life. The availability of solar energy has enabled the establishment of small businesses, improved educational facilities, and enhanced healthcare services. Moreover, the use of solar power aligns with Ladakh's cultural values of environmental stewardship and sustainability, preserving the region's pristine natural beauty for future generations.

In Australia, the town of Yackandandah is on a mission to become one of the first 100% renewable towns in the country. This ambitious goal is being driven by a grassroots community organization called Totally Renewable Yackandandah (TRY). The group has spearheaded efforts to install solar panels on homes, businesses, and public buildings, as well as develop a community-owned solar farm. The town's commitment to renewable energy is supported by innovative projects, such as the installation of battery storage systems and the creation of a virtual power plant that allows residents to share excess solar energy. Yackandandah's journey towards energy independence has not only reduced energy costs for residents but also fostered a strong sense of community and collaboration. The town's success has garnered national attention, inspiring other communities across Australia to pursue similar renewable energy goals.

In the Netherlands, the city of Utrecht is pioneering the integration of solar energy into urban infrastructure. The city's innovative approach includes the development of solar bike paths, which incorporate solar panels into the surface of cycling lanes. These solar paths generate clean electricity while providing a safe and sustainable mode of transportation for residents. Utrecht's commitment to renewable energy extends beyond solar bike paths, with initiatives to install solar panels on rooftops, parking lots, and public spaces throughout the city.

The integration of solar energy into urban planning not only reduces the city's carbon footprint but also enhances the quality of life for its residents by promoting sustainable mobility and green spaces.

These inspiring solar success stories demonstrate the transformative potential of renewable energy to drive positive change across diverse contexts. From rural villages to bustling cities, solar power is empowering communities, fostering economic growth, and promoting environmental sustainability. The lessons learned from these success stories offer valuable insights for others seeking to harness the power of the sun and create a brighter, more sustainable future. By sharing these stories, we can inspire action and collaboration, paving the way for a global transition to clean energy and a more resilient world.

Wind Energy Fundamentals

Harnessing the power of the wind has been a pursuit of humanity for centuries, from the sails of ancient ships to the windmills of the agricultural revolution. Today, wind energy stands as a pillar of the modern renewable energy landscape, offering a clean and sustainable alternative to fossil fuels. Understanding the fundamentals of wind energy involves exploring the science behind wind power, the technology used to capture it, and the factors that influence its efficiency and deployment.

At its core, wind energy is derived from the movement of air caused by the uneven heating of the Earth's surface by the sun. This movement creates kinetic energy, which can be captured and converted into electricity using wind turbines. The basic principle behind wind turbines is relatively simple: as the wind blows, it turns the blades of the turbine, which are connected to a rotor. The rotor spins a generator, producing electricity. The amount of energy generated depends on several factors, including wind speed, air density, and the size and design of the turbine.

Wind speed is a critical factor in determining the potential energy output of a wind turbine. The power generated by a turbine is proportional to the cube of the wind speed, meaning that even small increases in wind speed can lead to significant increases in energy production. This relationship underscores the importance of selecting optimal locations for wind farms, where wind speeds are consistently high. Coastal areas, open plains, and mountain passes are often ideal sites for wind energy projects due to their favorable wind conditions.

The design and size of wind turbines also play a crucial role in their efficiency and energy output. Modern wind turbines are engineering marvels, with blades that can reach lengths of over 80 meters and towers that soar to heights of 100 meters or more. These large-scale turbines are capable of capturing more wind energy and converting it into electricity more efficiently than their smaller counterparts. Advances in materials science and aerodynamics have led to the development of lighter, stronger, and more efficient turbine blades, further enhancing the performance of wind energy systems.

Wind turbines can be deployed in a variety of settings, from onshore wind farms to offshore installations. Onshore wind farms are typically located in rural areas with ample space and strong wind resources. These installations can range from small clusters of turbines to large-scale projects spanning hundreds of acres. Offshore wind farms, on the other hand, are situated in bodies of water, where wind speeds are generally higher and more consistent than on land. Offshore turbines are often larger and more powerful than onshore models, taking advantage of the open space and strong winds available at sea.

The integration of wind energy into the electrical grid presents both opportunities and challenges. Wind power is inherently variable, as it depends on the availability of wind, which can fluctuate over time. This variability requires careful management to ensure grid stability and reliability. Grid operators use a combination of forecasting, energy storage, and demand response strategies to balance the supply and demand of electricity and accommodate the intermittent nature of wind energy. Advances in grid technology, such as smart grids and advanced metering infrastructure, are enhancing the ability to integrate wind power into the energy mix effectively.

The environmental benefits of wind energy are significant, as it produces electricity without emitting greenhouse gases or other pollutants. By replacing fossil fuel-based power generation with wind energy, we can reduce carbon emissions and mitigate the impacts of climate change. Additionally, wind energy has a relatively small environmental footprint compared to other forms of energy production. While the construction and operation of wind farms can have localized impacts on wildlife and ecosystems, these effects are generally minimal and can be mitigated through careful planning and management.

Economically, wind energy offers a compelling case for investment and development. The cost of wind power has decreased dramatically over the past few decades, making it one of the most cost-competitive sources of electricity. This reduction in cost is driven by technological advancements, economies of scale, and increased competition in the wind energy industry. As a result, wind power is becoming an increasingly attractive option for utilities, businesses, and governments seeking to diversify their energy portfolios and reduce reliance on fossil fuels.

The growth of the wind energy sector is also a significant driver of job creation and economic development. The construction, operation, and maintenance of wind farms require a diverse workforce, providing employment opportunities across a range of skill levels and disciplines. From engineers and technicians to construction workers and environmental specialists, the wind energy industry supports a wide array of jobs and contributes to local and regional economies.

Public perception and acceptance of wind energy are important factors in its continued growth and development. While wind power is generally viewed favorably due to its environmental benefits, concerns about visual impact, noise, and effects on wildlife can pose challenges to project development. Engaging with communities and stakeholders early in the planning process is essential for addressing these concerns and building support for wind energy projects. Transparent communication, community involvement, and benefit-sharing mechanisms can help foster positive relationships and ensure the successful deployment of wind energy.

The future of wind energy is bright, with continued advancements in technology, policy support, and public engagement driving its expansion. As we strive to transition to a more sustainable energy system, wind power will play a crucial role in reducing carbon emissions, enhancing energy security, and supporting economic growth. By harnessing the power of the wind, we can create a cleaner, more resilient energy future for generations to come.

Technological Innovations in Wind Power

The evolution of wind power technology has been marked by a series of groundbreaking innovations that have significantly enhanced the efficiency, reliability, and scalability of wind energy systems. These technological advancements are driving the expansion of wind power as a key component of the global renewable energy portfolio, offering new opportunities for clean energy generation and integration.

One of the most notable innovations in wind power technology is the development of larger and more efficient wind turbines. Over the past few decades, the size of wind turbines has increased dramatically, with modern turbines boasting rotor diameters exceeding 160 meters and tower heights reaching over 200 meters. These larger turbines are capable of capturing more wind energy and converting it into electricity more efficiently than their smaller predecessors. The increase in size is complemented by advances in materials science, which have led to the creation of lighter and stronger turbine blades. These blades are designed to withstand the stresses of high wind speeds while maximizing energy capture, resulting in higher energy yields and improved performance.

The design of wind turbine blades has also seen significant innovation, with the introduction of advanced aerodynamic features that enhance efficiency. Modern blades often incorporate features such as serrated edges, winglets, and adaptive materials that adjust to changing wind conditions. These design elements reduce turbulence and drag, allowing turbines to operate more smoothly and efficiently. Additionally, the use of composite materials, such as carbon fiber and fiberglass, has improved the durability and longevity of turbine blades, reducing maintenance costs and extending the lifespan of wind energy systems.

Offshore wind technology represents another area of significant innovation in the wind power sector. Offshore wind farms are typically located in bodies of water, where wind speeds are generally higher and more consistent than on land. The development of floating wind turbines has opened up new possibilities for offshore wind energy, allowing turbines to be deployed in deeper waters where traditional fixed-bottom structures are not feasible. Floating turbines are anchored to the seabed using mooring lines, enabling them to harness the strong and steady winds found far from shore. This technology is expanding the potential for offshore wind energy, providing access to vast untapped wind resources and reducing the visual and environmental impact of nearshore installations.

The integration of digital technologies is revolutionizing the operation and management of wind energy systems. Advanced sensors and data analytics are being used to monitor the performance of wind turbines in real-time, enabling operators to optimize energy production and identify potential issues before they become critical. Predictive maintenance, powered by machine learning algorithms, allows for more efficient scheduling of maintenance activities, reducing downtime and extending the operational life of wind turbines. These digital innovations are enhancing the reliability and cost-effectiveness of wind energy, making it a more attractive option for utilities and investors.

Energy storage solutions are playing a crucial role in addressing the intermittency of wind power and enhancing grid integration. The development of high-capacity batteries and other storage technologies is enabling the storage of excess wind energy generated during periods of high wind, which can then be used during times of low wind or high demand. This capability not only improves the reliability of wind energy but also supports grid stability and flexibility. Hybrid systems that combine wind power with other renewable energy sources, such as solar or hydropower, are also being explored as a means of providing a more consistent and reliable energy supply.

The use of artificial intelligence and machine learning is further advancing the capabilities of wind energy systems. These technologies are being employed to improve wind forecasting, optimize turbine performance, and enhance grid integration. By analyzing vast amounts of data from weather models, turbine sensors, and grid operations, AI algorithms can provide more accurate predictions of wind energy generation and inform decision-making processes. This level of insight and control is enabling more efficient and effective management of wind energy resources, maximizing their contribution to the energy mix.

Innovations in grid infrastructure and management are also facilitating the integration of wind power into the energy system. The development of smart grids, which use digital communication technologies to monitor and manage energy flows, is enhancing the ability to accommodate the variable nature of wind energy. Smart grids enable more efficient energy distribution, demand response, and real-time monitoring, ensuring grid stability and reliability. Additionally, the expansion of transmission networks is allowing wind energy to be transported from remote locations with abundant wind resources to areas with high energy demand, further increasing the reach and impact of wind power.

The potential for wind energy to contribute to a sustainable energy future is immense, and continued technological innovation is key to unlocking this potential. By pushing the boundaries of what is possible with wind power, we can create more efficient, reliable, and scalable energy systems that meet the needs of a growing global population while reducing our reliance on fossil fuels. The journey of wind energy innovation is far from over, and the advancements yet to come hold the promise of even greater achievements in the quest for clean and sustainable energy.

Environmental and Economic Impacts

The shift towards renewable energy sources, particularly wind and solar power, is reshaping the environmental and economic landscapes in profound ways. As the world grapples with the dual challenges of climate change and energy security, understanding the impacts of these technologies is crucial for informed decision-making and sustainable development. The environmental and economic implications of renewable energy adoption are intertwined, offering both opportunities and challenges that must be navigated with care and foresight.

From an environmental perspective, the transition to renewable energy offers significant benefits in terms of reducing greenhouse gas emissions and mitigating climate change. Unlike fossil fuels, wind and solar power generate electricity without releasing carbon dioxide or other harmful pollutants into the atmosphere. This reduction in emissions is critical for meeting international climate targets and protecting ecosystems from the adverse effects of global warming. By replacing coal, oil, and natural gas with clean energy sources, we can decrease air pollution, improve public health, and preserve biodiversity.

The environmental footprint of renewable energy technologies is generally smaller than that of traditional energy sources. Wind and solar installations require less water for operation compared to fossil fuel power plants, which is particularly important in regions facing water scarcity. Additionally, the land use impact of renewable energy projects can be minimized through careful planning and site selection. For example, solar panels can be installed on rooftops or in areas with low ecological value, while wind turbines can be sited on agricultural land with minimal disruption to farming activities. By integrating renewable energy into existing landscapes, we can balance energy production with environmental conservation.

However, the deployment of renewable energy technologies is not without its environmental challenges. The manufacturing, transportation, and installation of wind turbines and solar panels require energy and resources, which can result in emissions and waste. The production of solar panels, in particular, involves the use of hazardous materials that must be managed responsibly to prevent environmental contamination. Additionally, the presence of wind turbines can impact local wildlife, particularly birds and bats, which may collide with turbine blades. Mitigating these impacts requires ongoing research, innovation, and collaboration between industry, government, and environmental organizations.

Economically, the transition to renewable energy presents both opportunities and challenges. One of the most significant economic benefits is the potential for job creation and economic growth. The renewable energy sector is a major driver of employment, supporting jobs in manufacturing, installation, maintenance, and research and development. As the demand for clean energy continues to grow, so too does the need for skilled workers across a range of disciplines. This growth not only supports local economies but also contributes to a more resilient and diversified workforce.

The cost of renewable energy has decreased dramatically over the past decade, making it increasingly competitive with fossil fuels. Advances in technology, economies of scale, and increased competition have driven down the cost of wind and solar power, enabling more widespread adoption. This reduction in cost is particularly beneficial for developing countries, where access to affordable and reliable energy is a key driver of economic development. By investing in renewable energy, these countries can reduce their dependence on imported fossil fuels, enhance energy security, and stimulate economic growth.

The integration of renewable energy into the grid presents economic challenges that must be addressed to ensure a stable and reliable energy supply. The variable nature of wind and solar power requires investment in grid infrastructure, energy storage, and demand response technologies to balance supply and demand. These investments can be costly, but they are essential for maximizing the benefits of renewable energy and ensuring grid stability. Policymakers and utilities must work together to develop regulatory frameworks and market mechanisms that support the integration of renewable energy while maintaining affordability and reliability for consumers.

The economic impacts of renewable energy extend beyond job creation and cost savings. The transition to clean energy can also drive innovation and technological advancement, creating new opportunities for businesses and entrepreneurs. The development of advanced materials, energy storage solutions, and smart grid technologies is opening up new markets and revenue streams, fostering a culture of innovation and collaboration. By embracing renewable energy, businesses can enhance their competitiveness, reduce their carbon footprint, and meet the growing demand for sustainable products and services.

Public perception and acceptance of renewable energy are important factors in its economic and environmental success. While renewable energy is generally viewed favorably due to its environmental benefits, concerns about visual impact, noise, and effects on local communities can pose challenges to project development. Engaging with stakeholders and communities early in the planning process is essential for addressing these concerns and building support for renewable energy projects. Transparent communication, community involvement, and benefit-sharing mechanisms can help foster positive relationships and ensure the successful deployment of renewable energy.

The transition to renewable energy is a complex and multifaceted process that requires coordinated efforts across various sectors and stakeholders. By understanding and addressing the environmental and economic impacts of renewable energy, we can accelerate the shift towards a sustainable energy future. The potential for renewable energy to transform our energy systems and improve quality of life is immense, and by overcoming these challenges, we can unlock its full potential and create a cleaner, more resilient world.

Community Engagement and Acceptance

Community engagement and acceptance are pivotal in the successful deployment of renewable energy projects. As the world transitions to cleaner energy sources, the role of local communities becomes increasingly significant. Their support can make or break a project, influencing its timeline, cost, and overall success. Understanding the dynamics of community engagement and fostering acceptance are essential for developers, policymakers, and stakeholders aiming to implement renewable energy solutions effectively.

The foundation of successful community engagement lies in transparent and open communication. From the outset, project developers must establish clear lines of communication with local residents, businesses, and community leaders. This involves sharing detailed information about the project's goals, benefits, potential impacts, and timelines. By providing stakeholders with a comprehensive understanding of the project, developers can build trust and address any concerns or misconceptions that may arise. Transparency not only fosters trust but also empowers communities to make informed decisions and actively participate in the planning process.

Listening to the concerns and aspirations of the community is equally important. Each community is unique, with its own set of values, priorities, and challenges. Engaging with community members through public meetings, workshops, and surveys allows developers to gain valuable insights into local perspectives and identify potential areas of conflict. By actively listening and responding to community feedback, developers can tailor their projects to better align with local needs and expectations. This collaborative approach not only enhances project design but also strengthens community support and ownership.

Incorporating community benefits into renewable energy projects is a powerful strategy for gaining acceptance and support. These benefits can take various forms, such as job creation, local investment, and revenue sharing. By demonstrating tangible economic and social benefits, developers can build goodwill and foster a sense of shared purpose. For example, hiring local workers for construction and maintenance, sourcing materials from local suppliers, and investing in community infrastructure can create a positive economic impact and strengthen community ties. Additionally, revenue-sharing agreements or community ownership models can provide long-term financial benefits, empowering communities to reinvest in local development initiatives.

Education and awareness-raising are critical components of community engagement. Many communities may have limited knowledge or experience with renewable energy technologies, leading to misconceptions or resistance. Providing educational resources, such as informational brochures, workshops, and site visits, can help demystify renewable energy and highlight its benefits. By showcasing successful case studies and sharing expert insights, developers can build confidence and enthusiasm for renewable energy projects. Education also plays a crucial role in addressing concerns related to environmental impact, health, and safety, ensuring that communities have a balanced and informed perspective.

Addressing environmental and social concerns is essential for gaining community acceptance. Renewable energy projects can have localized impacts on land use, wildlife, and cultural heritage, which may raise concerns among community members. Conducting thorough environmental and social impact assessments, and involving the community in these processes, can help identify and mitigate potential issues. By demonstrating a commitment to environmental stewardship and social responsibility, developers can build trust and credibility with the community. Implementing mitigation measures, such as habitat restoration, noise reduction, and visual impact minimization, can further alleviate concerns and enhance community support.

Building partnerships with local organizations and stakeholders is a valuable strategy for fostering community engagement. Collaborating with local governments, non-profit organizations, and educational institutions can provide additional resources, expertise, and credibility to renewable energy projects. These partnerships can facilitate community outreach, enhance project visibility, and strengthen local networks. By working together towards common goals, developers and community partners can create a more inclusive and supportive environment for renewable energy initiatives.

Flexibility and adaptability are key to successful community engagement. Projects may encounter unforeseen challenges or changes in community dynamics, requiring developers to adjust their strategies and approaches. Being responsive to community feedback and willing to make modifications to project plans can demonstrate a genuine commitment to collaboration and partnership. This adaptability not only enhances project outcomes but also reinforces community trust and confidence in the developer's intentions.

Celebrating milestones and successes is an important aspect of maintaining community engagement and enthusiasm. Recognizing and acknowledging the contributions of community members, partners, and stakeholders can foster a sense of pride and accomplishment. Hosting events, such as ribbon-cutting ceremonies, open houses, and community celebrations, provides opportunities to showcase project achievements and express gratitude for community support. These events also serve as platforms for ongoing dialogue and relationship-building, reinforcing the sense of community ownership and involvement.

The role of community engagement and acceptance in renewable energy projects cannot be overstated. By prioritizing transparent communication, active listening, and collaboration, developers can build strong relationships with communities and foster a supportive environment for renewable energy initiatives. The benefits of this approach extend beyond individual projects, contributing to a broader cultural shift towards sustainability and resilience. As communities become more engaged and informed, they are empowered to take an active role in shaping their energy futures, driving positive change for generations to come.

Future Prospects for Wind Energy

The horizon of wind energy is vast and promising, with future prospects that are set to redefine the global energy landscape. As the world intensifies its efforts to combat climate change and transition to sustainable energy systems, wind power emerges as a pivotal player in this transformation. The potential for wind energy to meet a significant portion of global electricity demand is immense, driven by technological advancements, policy support, and increasing public awareness.

One of the most exciting prospects for wind energy lies in the continued evolution of turbine technology. The trend towards larger and more efficient turbines is expected to persist, with innovations in materials and design pushing the boundaries of what is possible. Future turbines may feature even longer blades, optimized for capturing low-speed winds, and advanced materials that enhance durability and performance. These developments will enable wind farms to generate more electricity from the same amount of wind, increasing their capacity and efficiency.

Offshore wind energy is poised for significant growth, with the potential to unlock vast untapped wind resources. The development of floating wind turbines is a game-changer, allowing for the deployment of wind farms in deeper waters where traditional fixed-bottom structures are not feasible. This technology opens up new areas for wind energy development, particularly in regions with limited onshore wind resources. As floating wind technology matures and costs decrease, it is expected to play a crucial role in expanding the global wind energy capacity.

The integration of wind energy with other renewable energy sources and storage solutions is another promising avenue for future development. Hybrid systems that combine wind power with solar, hydropower, or battery storage can provide a more stable and reliable energy supply, addressing the intermittency challenges associated with wind energy. These integrated systems can optimize energy production, reduce costs, and enhance grid stability, making renewable energy more competitive with traditional fossil fuels.

Policy and regulatory frameworks will continue to shape the future of wind energy. Governments around the world are increasingly recognizing the importance of renewable energy in achieving climate goals and are implementing policies to support its growth. Incentives such as tax credits, feed-in tariffs, and renewable energy mandates are driving investment in wind energy projects. Additionally, international cooperation and agreements, such as the Paris Agreement, are fostering a global commitment to renewable energy development.

Public perception and acceptance of wind energy are expected to evolve as awareness of its benefits grows. As communities become more familiar with wind energy projects and their positive impacts, resistance is likely to decrease. Education and outreach efforts will play a crucial role in building public support and addressing concerns related to visual impact, noise, and wildlife. By engaging with communities and stakeholders, developers can foster a sense of ownership and pride in renewable energy projects, paving the way for broader acceptance and adoption.

The economic prospects for wind energy are also promising, with the potential for job creation and economic growth. The renewable energy sector is a major driver of employment, supporting jobs in manufacturing, installation, maintenance, and research and development. As the demand for clean energy continues to grow, so too does the need for skilled workers across a range of disciplines. This growth not only supports local economies but also contributes to a more resilient and diversified workforce.

Innovation and research will continue to drive the future of wind energy, with ongoing efforts to improve efficiency, reduce costs, and enhance performance. Advances in digital technologies, such as data analytics and artificial intelligence, are expected to play a significant role in optimizing wind energy systems. These technologies can improve wind forecasting, enhance turbine performance, and facilitate grid integration, maximizing the contribution of wind energy to the energy mix.

The potential for wind energy to contribute to a sustainable energy future is immense, and continued technological innovation is key to unlocking this potential. By pushing the boundaries of what is possible with wind power, we can create more efficient, reliable, and scalable energy systems that meet the needs of a growing global population while reducing our reliance on fossil fuels. The journey of wind energy innovation is far from over, and the advancements yet to come hold the promise of even greater achievements in the quest for clean and sustainable energy.

The future of wind energy is bright, with continued advancements in technology, policy support, and public engagement driving its expansion. As we strive to transition to a more sustainable energy system, wind power will play a crucial role in reducing carbon emissions, enhancing energy security, and supporting economic growth. By harnessing the power of the wind, we can create a cleaner, more resilient energy future for generations to come.

Chapter 4: Hydropower: The Force of Water

Basics of Hydropower Generation

Hydropower, one of the oldest and most reliable forms of renewable energy, harnesses the energy of flowing water to generate electricity. Its roots trace back to ancient civilizations that used water wheels for mechanical tasks, but today, hydropower stands as a cornerstone of modern energy systems, providing a significant portion of the world's electricity. Understanding the basics of hydropower generation involves exploring the principles behind it, the technology used, and the factors that influence its efficiency and deployment.

At its core, hydropower generation relies on the conversion of kinetic energy from moving water into mechanical energy, which is then transformed into electrical energy. This process typically occurs in a hydropower plant, where water flows through turbines, causing them to spin. The spinning turbines drive generators, which produce electricity. The amount of electricity generated depends on two main factors: the flow rate of the water and the height from which it falls, known as the head. The greater the flow and head, the more energy can be produced.

There are several types of hydropower plants, each suited to different geographical and hydrological conditions. The most common type is the impoundment facility, which uses a dam to store water in a reservoir. By controlling the release of water, these plants can generate electricity on demand, making them highly flexible and reliable. Run-of-river plants, on the other hand, do not require large reservoirs. Instead, they divert a portion of a river's flow through a canal or penstock to generate electricity. These plants have a smaller environmental footprint but are more dependent on natural water flow variations.

Pumped storage hydropower is a unique type of facility that acts as a large-scale energy storage system. During periods of low electricity demand, excess energy is used to pump water from a lower reservoir to an upper reservoir. When demand increases, the stored water is released back to the lower reservoir, passing through turbines to generate electricity. This ability to store and release energy makes pumped storage an essential component of modern energy grids, providing stability and flexibility.

The environmental impact of hydropower is generally lower than that of fossil fuel-based energy sources, but it is not without challenges. The construction of dams and reservoirs can disrupt local ecosystems, affecting fish populations and altering natural water flow patterns. To mitigate these impacts, modern hydropower projects often incorporate fish ladders, bypass systems, and other measures to support aquatic life. Additionally, careful site selection and environmental assessments are crucial to minimizing the ecological footprint of hydropower facilities.

Hydropower offers several advantages that make it an attractive option for electricity generation. It is a renewable resource, as it relies on the natural water cycle, and it produces electricity without emitting greenhouse gases or other pollutants. Hydropower plants also have long lifespans, often exceeding 50 years, and relatively low operating and maintenance costs. Their ability to provide baseload power and respond quickly to changes in electricity demand makes them valuable assets for grid stability and reliability.

The economic benefits of hydropower extend beyond electricity generation. The construction and operation of hydropower plants create jobs and stimulate local economies. In many cases, reservoirs associated with hydropower projects provide additional benefits, such as water supply for irrigation, flood control, and recreational opportunities. These multifaceted advantages contribute to the overall value of hydropower as a sustainable energy solution.

Despite its benefits, the development of new hydropower projects faces several challenges. Environmental concerns, regulatory hurdles, and social opposition can complicate project planning and implementation. Additionally, the availability of suitable sites for large-scale hydropower development is limited, particularly in regions with dense populations or competing land uses. As a result, the focus is increasingly shifting towards optimizing existing facilities and exploring innovative technologies, such as small-scale and micro-hydropower systems.

Small-scale hydropower systems, often referred to as micro-hydro, offer a promising solution for remote or off-grid communities. These systems typically generate less than 10 megawatts of electricity and can be installed in small rivers or streams with minimal environmental impact. By providing a reliable and sustainable source of energy, micro-hydro systems can support rural development, improve quality of life, and reduce reliance on fossil fuels.

The integration of digital technologies is enhancing the efficiency and performance of hydropower systems. Advanced monitoring and control systems enable operators to optimize water flow, turbine performance, and energy output. Predictive maintenance, powered by data analytics, allows for timely identification and resolution of potential issues, reducing downtime and extending the lifespan of equipment. These innovations are contributing to the modernization of hydropower facilities and their continued relevance in the evolving energy landscape.

The future of hydropower lies in balancing its benefits with environmental and social considerations. By adopting best practices in design, construction, and operation, hydropower projects can minimize their ecological footprint and maximize their contributions to sustainable development. Collaboration between governments, industry, and communities is essential to address challenges and unlock the full potential of hydropower as a clean and reliable energy source.

As the world seeks to transition to a more sustainable energy system, hydropower will continue to play a vital role in meeting electricity demand and supporting economic growth. Its ability to provide renewable, low-emission energy, coupled with its flexibility and reliability, makes it an indispensable component of the global energy mix. By understanding the basics of hydropower generation and embracing innovation, we can harness the power of water to create a cleaner, more resilient energy future.

Types and Applications of Hydropower

Hydropower, a versatile and time-tested source of renewable energy, comes in various forms, each tailored to specific geographical and hydrological conditions. Understanding the different types of hydropower and their applications is essential for harnessing the full potential of this energy source. From large-scale dams to small community projects, hydropower offers a range of solutions to meet diverse energy needs while contributing to sustainable development.

The most common type of hydropower is the impoundment facility, which involves the construction of a dam to create a reservoir. This type of hydropower plant is designed to store water and release it as needed to generate electricity. The reservoir acts as a buffer, allowing for the regulation of water flow and the production of electricity on demand. Impoundment facilities are capable of generating large amounts of electricity, making them suitable for meeting baseload power requirements. They also offer additional benefits, such as flood control, irrigation, and recreational opportunities.

Run-of-river hydropower is another widely used type of hydropower, characterized by its minimal impact on river flow and ecosystems. Unlike impoundment facilities, run-of-river plants do not require large reservoirs. Instead, they divert a portion of a river's flow through a canal or penstock to generate electricity. This type of hydropower is particularly well-suited for regions with consistent river flow and limited space for reservoirs. Run-of-river plants have a smaller environmental footprint and can be integrated into existing river systems with minimal disruption.

Pumped storage hydropower is a unique type of facility that serves as both a power generator and an energy storage system. During periods of low electricity demand, excess energy is used to pump water from a lower reservoir to an upper reservoir. When demand increases, the stored water is released back to the lower reservoir, passing through turbines to generate electricity. This ability to store and release energy makes pumped storage an essential component of modern energy grids, providing stability and flexibility. It is particularly valuable for balancing intermittent renewable energy sources, such as wind and solar power.

Small-scale hydropower, often referred to as micro-hydro, offers a promising solution for remote or off-grid communities. These systems typically generate less than 10 megawatts of electricity and can be installed in small rivers or streams with minimal environmental impact. Micro-hydro systems are highly adaptable and can be tailored to meet the specific needs of local communities. By providing a reliable and sustainable source of energy, micro-hydro systems can support rural development, improve quality of life, and reduce reliance on fossil fuels.

The applications of hydropower extend beyond electricity generation, offering a range of benefits that contribute to economic and social development. In many regions, hydropower projects provide water supply for irrigation, supporting agriculture and food security. The reservoirs associated with hydropower facilities can also serve as sources of drinking water, enhancing access to clean water for local communities. Additionally, hydropower projects often create opportunities for recreation and tourism, such as boating, fishing, and hiking, which can stimulate local economies and promote environmental conservation.

Hydropower's role in flood control is another important application, particularly in regions prone to seasonal flooding.

By regulating water flow and storing excess water during periods of heavy rainfall, hydropower dams can reduce the risk of flooding and protect downstream communities. This capability not only safeguards lives and property but also supports economic stability by minimizing disruptions to agriculture, infrastructure, and industry.

The integration of hydropower with other renewable energy sources is an emerging trend that offers new opportunities for enhancing energy systems. Hybrid systems that combine hydropower with solar, wind, or biomass energy can provide a more stable and reliable energy supply, addressing the intermittency challenges associated with renewable energy. These integrated systems can optimize energy production, reduce costs, and enhance grid stability, making renewable energy more competitive with traditional fossil fuels.

Despite its many advantages, the development of hydropower projects must be carefully managed to balance environmental and social considerations. The construction of dams and reservoirs can disrupt local ecosystems, affecting fish populations and altering natural water flow patterns. To mitigate these impacts, modern hydropower projects often incorporate fish ladders, bypass systems, and other measures to support aquatic life. Additionally, careful site selection and environmental assessments are crucial to minimizing the ecological footprint of hydropower facilities.

Community engagement and stakeholder involvement are essential for the successful implementation of hydropower projects. By involving local communities in the planning and decision-making process, developers can build trust, address concerns, and foster a sense of ownership. Transparent communication, benefit-sharing mechanisms, and capacity-building initiatives can enhance community support and ensure that hydropower projects contribute to local development goals.

Innovation and research continue to drive the evolution of hydropower technology, with ongoing efforts to improve efficiency, reduce costs, and enhance performance. Advances in turbine design, materials, and digital technologies are contributing to the modernization of hydropower facilities and their continued relevance in the evolving energy landscape. By embracing innovation and adopting best practices, the hydropower sector can unlock new opportunities for sustainable energy generation and support the transition to a low-carbon future.

Hydropower remains a vital component of the global energy mix, offering a reliable, renewable, and versatile source of electricity. Its ability to provide baseload power, support grid stability, and deliver a range of economic and social benefits makes it an indispensable part of the transition to sustainable energy systems. By understanding the types and applications of hydropower, we can harness its potential to create a cleaner, more resilient energy future for generations to come.

Balancing Environmental Concerns

Balancing environmental concerns with the development of renewable energy projects is a complex yet essential task. As the world increasingly turns to sustainable energy sources to combat climate change, it is crucial to ensure that these projects do not inadvertently harm the ecosystems they aim to protect. This delicate balance requires careful planning, innovative solutions, and a commitment to environmental stewardship.

The first step in addressing environmental concerns is conducting thorough environmental impact assessments (EIAs). These assessments evaluate the potential effects of a project on local ecosystems, wildlife, and communities. By identifying potential risks and impacts early in the planning process, developers can design projects that minimize harm and enhance environmental benefits. EIAs also provide a platform for engaging with stakeholders, including local communities, environmental organizations, and government agencies, to gather input and address concerns.

Mitigating the impact on wildlife is a significant consideration for renewable energy projects, particularly wind and solar installations. Wind turbines, for example, can pose risks to birds and bats, which may collide with the blades. To reduce these risks, developers can implement measures such as siting turbines away from migratory paths, using radar technology to detect approaching birds, and adjusting turbine operation during peak migration periods. Similarly, solar farms can impact local habitats, but careful site selection and habitat restoration efforts can mitigate these effects.

Hydropower projects, while offering substantial renewable energy benefits, can also have significant environmental impacts, particularly on aquatic ecosystems. The construction of dams and reservoirs can alter natural water flow, affect fish populations, and disrupt sediment transport. To address these concerns, modern hydropower projects often incorporate fish ladders, bypass systems, and sediment management strategies.

These measures help maintain ecological connectivity and support the health of aquatic ecosystems.

The siting of renewable energy projects is a critical factor in balancing environmental concerns. Selecting locations with low ecological sensitivity can significantly reduce the environmental footprint of a project. For instance, installing solar panels on rooftops or in brownfield sites can avoid the need for land conversion and habitat disruption. Similarly, offshore wind farms can be sited in areas with minimal impact on marine life and shipping routes. By prioritizing environmentally responsible siting, developers can minimize conflicts and enhance the sustainability of their projects.

Community engagement plays a vital role in addressing environmental concerns and building support for renewable energy projects. By involving local communities in the planning and decision-making process, developers can gain valuable insights into local environmental priorities and concerns. Transparent communication, participatory planning, and benefit-sharing mechanisms can foster trust and collaboration, ensuring that projects align with community values and contribute to local development goals.

Innovation and technology are key drivers in balancing environmental concerns with renewable energy development. Advances in materials, design, and monitoring systems are enabling more efficient and environmentally friendly energy solutions. For example, the development of bird-friendly turbine designs, bifacial solar panels, and floating solar arrays are helping to reduce the environmental impact of renewable energy projects. Additionally, digital technologies, such as remote sensing and data analytics, are enhancing the ability to monitor and manage environmental impacts in real-time.

Policy and regulatory frameworks play a crucial role in guiding the development of renewable energy projects in an environmentally responsible manner. Governments can establish clear guidelines and standards for environmental protection, incentivize best practices, and support research and innovation. By creating a supportive policy environment, governments can encourage the adoption of sustainable energy solutions that balance environmental, social, and economic considerations.

The integration of renewable energy with conservation efforts offers a promising pathway for enhancing environmental outcomes. By aligning renewable energy projects with conservation goals, such as habitat restoration, reforestation, and biodiversity protection, developers can create synergies that benefit both energy production and ecosystem health. Collaborative partnerships between energy companies, conservation organizations, and local communities can drive innovative solutions and maximize the positive impact of renewable energy projects.

Adaptive management is an essential approach for balancing environmental concerns in renewable energy development. This approach involves continuously monitoring and evaluating the environmental impacts of a project and making adjustments as needed to minimize harm and enhance benefits. By embracing flexibility and learning from experience, developers can improve the environmental performance of their projects over time and respond effectively to changing conditions and new information.

The transition to renewable energy presents an opportunity to redefine the relationship between energy production and environmental stewardship. By prioritizing sustainability, innovation, and collaboration, the renewable energy sector can lead the way in creating a more harmonious balance between human needs and the natural world. As we strive to meet global energy demands while protecting the planet, the lessons learned from balancing environmental concerns will be invaluable in shaping a sustainable energy future.

The journey towards sustainable energy is a shared responsibility that requires the collective efforts of governments, industry, communities, and individuals. By working together to address environmental concerns and embrace innovative solutions, we can unlock the full potential of renewable energy and create a cleaner, more resilient world for future generations.

Economic Viability and Benefits

Economic viability is a cornerstone of any successful renewable energy project, and understanding the financial dynamics of such initiatives is crucial for stakeholders, investors, and policymakers. The transition to renewable energy sources, such as wind, solar, and hydropower, is not only driven by environmental imperatives but also by the potential for economic growth and stability. The benefits of renewable energy extend beyond mere cost savings, encompassing job creation, energy security, and long-term sustainability.

The initial investment in renewable energy projects can be substantial, often requiring significant capital outlay for technology, infrastructure, and development. However, the long-term economic benefits often outweigh these initial costs. Renewable energy sources, once operational, typically have lower operating and maintenance costs compared to fossil fuel-based systems. This is largely due to the absence of fuel costs, as wind, sunlight, and water are free resources. Over time, these savings can lead to a favorable return on investment, making renewable energy projects economically attractive.

One of the key factors influencing the economic viability of renewable energy is the levelized cost of electricity (LCOE). LCOE represents the average cost of generating electricity over the lifetime of a project, taking into account capital, operating, and maintenance expenses. In recent years, the LCOE for renewable energy technologies has decreased significantly, driven by technological advancements, economies of scale, and increased competition. As a result, renewables are becoming increasingly competitive with, and in some cases cheaper than, traditional fossil fuels.

Government policies and incentives play a crucial role in enhancing the economic viability of renewable energy projects. Tax credits, grants, and subsidies can reduce the financial burden on developers and encourage investment in clean energy. Feed-in tariffs and renewable energy certificates provide additional revenue streams, further improving the financial outlook for renewable projects. By creating a supportive policy environment, governments can accelerate the transition to renewable energy and stimulate economic growth.

The economic benefits of renewable energy extend beyond cost savings and financial returns. The renewable energy sector is a major driver of job creation, supporting employment across a range of industries, including manufacturing, construction, installation, and maintenance. As the demand for clean energy continues to grow, so too does the need for skilled workers, creating opportunities for workforce development and training. This growth not only supports local economies but also contributes to a more resilient and diversified workforce.

Energy security is another significant economic benefit of renewable energy. By reducing reliance on imported fossil fuels, countries can enhance their energy independence and reduce vulnerability to global energy market fluctuations. This stability can lead to more predictable energy prices, benefiting consumers and businesses alike. Additionally, the decentralized nature of many renewable energy systems, such as rooftop solar panels and community wind projects, can enhance grid resilience and reduce the risk of power outages.

Renewable energy projects can also stimulate local economic development by providing additional revenue streams for communities. For example, wind farms and solar installations often involve land lease agreements that provide income to landowners. Community ownership models and revenue-sharing agreements can further distribute economic benefits, empowering communities to reinvest in local development initiatives. These arrangements can foster a sense of ownership and pride, enhancing community support for renewable energy projects.

The environmental benefits of renewable energy, while often discussed in terms of sustainability, also have economic implications. By reducing greenhouse gas emissions and air pollution, renewable energy can lead to improved public health outcomes and reduced healthcare costs. Cleaner air and water contribute to a healthier population, reducing the economic burden of pollution-related illnesses and enhancing overall quality of life.

Innovation and technological advancements continue to drive the economic viability of renewable energy. Research and development efforts are focused on improving efficiency, reducing costs, and enhancing performance. Advances in energy storage, grid integration, and digital technologies are enabling more efficient and flexible energy systems, maximizing the economic benefits of renewable energy. By embracing innovation, the renewable energy sector can unlock new opportunities for growth and competitiveness.

The transition to renewable energy presents an opportunity to redefine economic growth in a sustainable and inclusive manner. By prioritizing renewable energy development, countries can create a more resilient and equitable economy that benefits all stakeholders. The economic viability of renewable energy is not just about financial returns; it is about creating a sustainable future that balances economic, environmental, and social considerations.

The journey towards a renewable energy future is a shared responsibility that requires collaboration and commitment from governments, industry, communities, and individuals. By working together to address economic challenges and embrace innovative solutions, we can unlock the full potential of renewable energy and create a cleaner, more prosperous world for future generations.

Innovations in Hydropower Technology

Hydropower, a stalwart of renewable energy, is undergoing a transformation driven by technological innovation. As the world seeks to expand its renewable energy portfolio, advancements in hydropower technology are unlocking new possibilities for efficiency, sustainability, and integration. These innovations are not only enhancing the performance of existing hydropower systems but also paving the way for novel applications and solutions that address contemporary energy challenges.

One of the most significant areas of innovation in hydropower technology is turbine design. Traditional turbines, while effective, are being reimagined to improve efficiency and reduce environmental impact. Advanced turbine designs, such as fish-friendly turbines, are engineered to minimize harm to aquatic life by allowing fish to pass safely through the system. These turbines incorporate features like smoother surfaces and optimized blade shapes to reduce injury and mortality rates among fish populations. By addressing ecological concerns, these innovations make hydropower a more sustainable option for energy generation.

Variable speed turbines represent another leap forward in turbine technology. Unlike conventional turbines that operate at a fixed speed, variable speed turbines can adjust their rotational speed to match the flow of water. This flexibility allows for more efficient energy capture, particularly in environments with fluctuating water flow. By optimizing energy production under varying conditions, variable speed turbines enhance the overall efficiency and reliability of hydropower plants.

The integration of digital technologies is revolutionizing the operation and management of hydropower facilities. Advanced monitoring and control systems, powered by data analytics, enable operators to optimize water flow, turbine performance, and energy output. Real-time data collection and analysis provide insights into system performance, allowing for predictive maintenance and timely identification of potential issues. This proactive approach reduces downtime, extends equipment lifespan, and enhances the overall efficiency of hydropower operations.

Energy storage solutions are playing an increasingly important role in the evolution of hydropower technology. Pumped storage hydropower, a well-established form of energy storage, is being enhanced with new technologies that improve efficiency and flexibility. Innovations such as closed-loop systems, which operate independently of natural water bodies, offer greater control over water flow and reduce environmental impact. These systems can be strategically located to complement other renewable energy sources, such as wind and solar, providing grid stability and balancing intermittent energy supply.

Small-scale and micro-hydropower systems are gaining traction as versatile solutions for decentralized energy generation. These systems, often installed in remote or off-grid locations, provide reliable and sustainable energy to communities with limited access to traditional power sources. Innovations in micro-hydro technology, such as modular and portable systems, are making it easier to deploy and maintain these installations. By harnessing local water resources, micro-hydropower systems empower communities to achieve energy independence and support local development.

Floating hydropower is an emerging technology that offers new opportunities for energy generation in areas with limited land availability. By installing hydropower systems on floating platforms, developers can tap into water bodies such as reservoirs, lakes, and even oceans. This approach minimizes land use and environmental disruption while providing a stable platform for energy production. Floating hydropower can be integrated with other renewable energy technologies, such as floating solar panels, to create hybrid systems that maximize energy output and efficiency.

The use of advanced materials is another area of innovation in hydropower technology. New materials, such as composite and corrosion-resistant alloys, are being used to construct turbines and other components, enhancing durability and performance. These materials reduce wear and tear, lower maintenance costs, and extend the lifespan of hydropower systems. By improving the resilience of hydropower infrastructure, these innovations contribute to the long-term sustainability and economic viability of hydropower projects.

Environmental monitoring and mitigation technologies are essential for ensuring the sustainability of hydropower projects. Innovations in remote sensing, satellite imagery, and environmental modeling provide valuable data on ecosystem health and project impact. These technologies enable developers to assess and mitigate potential environmental risks, such as changes in water quality, sediment transport, and habitat disruption. By incorporating environmental considerations into project design and operation, hydropower developers can balance energy production with ecological preservation.

Collaboration and knowledge sharing are driving the advancement of hydropower technology. Partnerships between research institutions, industry, and government agencies are fostering innovation and accelerating the development of new solutions. Collaborative initiatives, such as international research programs and technology demonstration projects, provide platforms for testing and refining new technologies. By leveraging collective expertise and resources, the hydropower sector can overcome challenges and unlock new opportunities for growth and sustainability.

The future of hydropower technology is bright, with ongoing research and development efforts poised to deliver even greater advancements. As the world transitions to a more sustainable energy system, hydropower will continue to play a vital role in meeting global energy needs. By embracing innovation and adopting cutting-edge technologies, the hydropower sector can enhance its contribution to a cleaner, more resilient energy future.

The journey of hydropower innovation is far from over, and the advancements yet to come hold the promise of even greater achievements in the quest for sustainable energy. By harnessing the power of water and technology, we can create more efficient, reliable, and scalable energy systems that meet the needs of a growing global population while reducing our reliance on fossil fuels. The potential for hydropower to contribute to a sustainable energy future is immense, and continued technological innovation is key to unlocking this potential.

Chapter 5: Biomass and Bioenergy: Nature's Recyclers

Understanding Biomass Energy

Biomass energy, a cornerstone of renewable energy solutions, harnesses the power of organic materials to produce heat, electricity, and biofuels. This versatile energy source is derived from plant and animal matter, including wood, agricultural residues, and even municipal waste. Understanding the intricacies of biomass energy is essential for leveraging its potential to contribute to a sustainable energy future.

At its core, biomass energy is rooted in the natural process of photosynthesis, where plants convert sunlight into chemical energy stored in their tissues. When these organic materials are burned or processed, the stored energy is released, providing a renewable source of power. Unlike fossil fuels, which release carbon dioxide that has been sequestered for millions of years, biomass energy is considered carbon-neutral. This is because the carbon dioxide emitted during combustion is offset by the carbon dioxide absorbed by plants during their growth cycle.

One of the most common forms of biomass energy is direct combustion, where organic materials are burned to produce heat. This heat can be used directly for industrial processes, space heating, or to generate electricity through steam turbines. Wood, agricultural residues, and dedicated energy crops are often used as feedstocks for direct combustion. Advances in combustion technology have improved the efficiency and emissions profile of biomass power plants, making them a viable alternative to coal-fired power stations.

Biogas production is another important application of biomass energy. Through a process called anaerobic digestion, organic materials such as animal manure, food waste, and sewage are broken down by microorganisms in the absence of oxygen. This process produces biogas, a mixture of methane and carbon dioxide, which can be used as a renewable fuel for heating, electricity generation, or as a vehicle fuel. Biogas systems offer the added benefit of waste management, reducing the environmental impact of organic waste disposal.

Liquid biofuels, such as ethanol and biodiesel, represent a significant segment of the biomass energy landscape. Ethanol is typically produced from sugar- and starch-rich crops like corn and sugarcane through fermentation. It is commonly used as a gasoline additive to reduce emissions and enhance fuel efficiency. Biodiesel, on the other hand, is produced from vegetable oils, animal fats, or recycled cooking oils through a process called transesterification. Biodiesel can be used in diesel engines with little or no modification, offering a renewable alternative to petroleum-based diesel.

The versatility of biomass energy extends to its ability to integrate with existing energy systems. Co-firing, for example, involves burning biomass alongside coal in power plants, reducing greenhouse gas emissions and extending the life of existing infrastructure. This approach allows for a gradual transition to renewable energy while maintaining energy security and reliability. Additionally, biomass energy can be used in combined heat and power (CHP) systems, which simultaneously produce electricity and useful heat, maximizing energy efficiency.

Sustainability is a key consideration in the development and deployment of biomass energy. The sourcing of biomass feedstocks must be managed carefully to avoid negative environmental impacts, such as deforestation, soil degradation, and loss of biodiversity. Sustainable biomass production involves practices like reforestation, crop rotation, and the use of agricultural residues, which help maintain ecosystem health and productivity. Certification schemes and sustainability standards play a crucial role in ensuring that biomass energy contributes positively to environmental and social goals.

Economic viability is another important aspect of biomass energy. The cost of biomass feedstocks, transportation, and processing can vary significantly depending on local conditions and market dynamics. However, the potential for job creation and rural development makes biomass energy an attractive option for many regions. By providing a market for agricultural residues and waste materials, biomass energy can support local economies and enhance energy security.

Technological innovation continues to drive the evolution of biomass energy. Advances in conversion technologies, such as gasification and pyrolysis, are expanding the range of feedstocks that can be used and improving the efficiency of biomass energy systems. Gasification involves converting biomass into a combustible gas mixture, which can be used for electricity generation or as a feedstock for chemical production. Pyrolysis, on the other hand, involves heating biomass in the absence of oxygen to produce bio-oil, which can be refined into transportation fuels or used as a chemical feedstock.

The integration of biomass energy with other renewable energy sources offers new opportunities for enhancing energy systems.

Hybrid systems that combine biomass with solar, wind, or hydropower can provide a more stable and reliable energy supply, addressing the intermittency challenges associated with renewable energy. These integrated systems can optimize energy production, reduce costs, and enhance grid stability, making renewable energy more competitive with traditional fossil fuels.

Community engagement and stakeholder involvement are essential for the successful implementation of biomass energy projects. By involving local communities in the planning and decision-making process, developers can build trust, address concerns, and foster a sense of ownership. Transparent communication, benefit-sharing mechanisms, and capacity-building initiatives can enhance community support and ensure that biomass energy projects contribute to local development goals.

The future of biomass energy is bright, with ongoing research and development efforts poised to deliver even greater advancements. As the world transitions to a more sustainable energy system, biomass energy will continue to play a vital role in meeting global energy needs. By embracing innovation and adopting best practices, the biomass energy sector can enhance its contribution to a cleaner, more resilient energy future.

The journey towards sustainable energy is a shared responsibility that requires the collective efforts of governments, industry, communities, and individuals. By working together to address environmental and economic challenges and embrace innovative solutions, we can unlock the full potential of biomass energy and create a cleaner, more prosperous world for future generations.

Conversion Processes and Technologies

Conversion processes and technologies form the backbone of transforming raw energy sources into usable power, playing a pivotal role in the renewable energy landscape. These processes are the bridge between natural resources and the energy that powers our homes, industries, and transportation systems. Understanding the intricacies of these technologies is essential for anyone looking to delve into the world of renewable energy.

At the heart of conversion processes is the transformation of energy from one form to another. This can involve converting sunlight into electricity, biomass into biofuels, or wind into mechanical power. Each conversion process is unique, with its own set of technologies, efficiencies, and challenges. The choice of technology often depends on the type of energy source, the intended application, and the specific requirements of the energy system.

Photovoltaic (PV) technology is a prime example of a conversion process that transforms sunlight directly into electricity. PV cells, commonly known as solar cells, are made from semiconductor materials that exhibit the photovoltaic effect. When sunlight strikes these cells, it excites electrons, creating an electric current. This direct conversion of light into electricity is a clean and efficient way to harness solar energy. Advances in PV technology, such as the development of thin-film solar cells and multi-junction cells, are continually improving efficiency and reducing costs, making solar power more accessible and widespread.

Wind energy conversion involves transforming the kinetic energy of wind into mechanical power, which is then converted into electricity. Wind turbines, the primary technology for this conversion, consist of blades that capture wind energy and a generator that produces electricity. The design and placement of wind turbines are critical factors in maximizing energy capture and efficiency. Innovations in turbine design, such as larger rotor diameters and taller towers, are enhancing the performance of wind energy systems, allowing them to generate more power even in low-wind conditions.

Biomass conversion processes are diverse, encompassing a range of technologies that transform organic materials into heat, electricity, or biofuels. Combustion is the most straightforward method, where biomass is burned to produce heat and power. However, more advanced technologies, such as gasification and pyrolysis, offer greater efficiency and flexibility. Gasification involves converting biomass into a combustible gas mixture, which can be used for electricity generation or as a feedstock for chemical production. Pyrolysis, on the other hand, involves heating biomass in the absence of oxygen to produce bio-oil, which can be refined into transportation fuels or used as a chemical feedstock.

Anaerobic digestion is another important biomass conversion process, where microorganisms break down organic materials in the absence of oxygen to produce biogas. This biogas, primarily composed of methane and carbon dioxide, can be used as a renewable fuel for heating, electricity generation, or as a vehicle fuel. Anaerobic digestion not only provides a sustainable energy source but also offers waste management benefits by reducing the environmental impact of organic waste disposal.

Hydropower conversion processes harness the energy of flowing or falling water to generate electricity. This is typically achieved through the use of turbines and generators in hydropower plants. The potential energy of water stored in reservoirs or the kinetic energy of flowing rivers is converted into mechanical energy by the turbines, which is then transformed into electricity by the generators. Advances in turbine technology, such as fish-friendly designs and variable speed turbines, are improving the efficiency and environmental performance of hydropower systems.

Geothermal energy conversion involves tapping into the Earth's internal heat to produce electricity or provide direct heating. Geothermal power plants use steam or hot water from underground reservoirs to drive turbines and generate electricity. Direct-use applications involve using geothermal heat for industrial processes, greenhouse heating, or district heating systems. Enhanced geothermal systems (EGS) are an emerging technology that involves creating artificial reservoirs to extract heat from dry rock formations, expanding the potential for geothermal energy in regions without natural geothermal resources.

The integration of energy storage technologies with conversion processes is becoming increasingly important in the renewable energy sector. Energy storage systems, such as batteries, pumped hydro storage, and thermal storage, allow for the capture and storage of excess energy generated during periods of high production. This stored energy can then be released during periods of low production or high demand, enhancing the reliability and stability of renewable energy systems. The development of advanced storage technologies, such as lithium-ion batteries and flow batteries, is enabling more efficient and cost-effective energy storage solutions.

The efficiency of conversion processes is a critical factor in the overall performance of renewable energy systems. Improving conversion efficiency reduces energy losses and maximizes the output from available resources. Research and development efforts are focused on enhancing the efficiency of existing technologies and developing new materials and methods to improve energy conversion. Innovations such as perovskite solar cells, superconducting wind turbine generators, and advanced biofuel catalysts are pushing the boundaries of what is possible in energy conversion.

The environmental impact of conversion processes is another important consideration. While renewable energy technologies are generally more environmentally friendly than fossil fuels, they can still have ecological and social impacts. Sustainable practices, such as responsible sourcing of materials, minimizing land use, and reducing emissions, are essential for ensuring that conversion processes contribute positively to environmental and social goals. Certification schemes and sustainability standards play a crucial role in promoting best practices and ensuring the responsible development of renewable energy technologies.

The future of conversion processes and technologies is bright, with ongoing research and development efforts poised to deliver even greater advancements. As the world transitions to a more sustainable energy system, these technologies will continue to play a vital role in meeting global energy needs. By embracing innovation and adopting best practices, the renewable energy sector can enhance its contribution to a cleaner, more resilient energy future.

The journey towards sustainable energy is a shared responsibility that requires the collective efforts of governments, industry, communities, and individuals. By working together to address environmental and economic challenges and embrace innovative solutions, we can unlock the full potential of conversion processes and technologies and create a cleaner, more prosperous world for future generations.

Environmental and Economic Considerations

Balancing environmental and economic considerations is crucial in the development and deployment of renewable energy projects. As the world shifts towards sustainable energy solutions, understanding the interplay between ecological impacts and financial viability becomes essential. This chapter delves into the complexities of these considerations, offering insights into how they shape the renewable energy landscape.

Renewable energy projects are often lauded for their potential to reduce greenhouse gas emissions and mitigate climate change. However, they are not without their environmental impacts. The construction and operation of renewable energy facilities can affect local ecosystems, wildlife, and land use. For instance, large-scale solar farms may require significant land area, potentially disrupting habitats and altering land use patterns. Similarly, wind farms can pose risks to bird and bat populations if not carefully sited and managed.

To address these challenges, developers must conduct thorough environmental impact assessments (EIAs) before embarking on renewable energy projects. EIAs evaluate the potential ecological effects of a project and propose mitigation measures to minimize harm. These assessments consider factors such as habitat disruption, water usage, and emissions, ensuring that projects align with environmental regulations and sustainability goals. By integrating environmental considerations into project planning and design, developers can reduce negative impacts and enhance the overall sustainability of renewable energy initiatives.

Economic considerations are equally important in the renewable energy sector. The financial viability of a project determines its feasibility and long-term success. Initial capital costs, operating expenses, and potential revenue streams all play a role in shaping the economic outlook of renewable energy projects. While the upfront investment for renewable technologies can be substantial, the long-term benefits often outweigh these costs. Renewable energy sources typically have lower operating and maintenance expenses compared to fossil fuels, as they rely on free resources like sunlight and wind.

Government policies and incentives are critical in enhancing the economic attractiveness of renewable energy projects. Tax credits, grants, and subsidies can reduce the financial burden on developers and encourage investment in clean energy. Feed-in tariffs and renewable energy certificates provide additional revenue streams, further improving the financial outlook for renewable projects. By creating a supportive policy environment, governments can accelerate the transition to renewable energy and stimulate economic growth.

Job creation is another significant economic benefit of renewable energy. The sector supports employment across a range of industries, including manufacturing, construction, installation, and maintenance. As the demand for clean energy continues to grow, so too does the need for skilled workers, creating opportunities for workforce development and training. This growth not only supports local economies but also contributes to a more resilient and diversified workforce.

Energy security is a key consideration in both environmental and economic contexts. By reducing reliance on imported fossil fuels, countries can enhance their energy independence and reduce vulnerability to global energy market fluctuations. This stability can lead to more predictable energy prices, benefiting consumers and businesses alike. Additionally, the decentralized nature of many renewable energy systems, such as rooftop solar panels and community wind projects, can enhance grid resilience and reduce the risk of power outages.

The integration of renewable energy with existing energy systems presents both challenges and opportunities. On one hand, the intermittent nature of some renewable sources, such as solar and wind, can pose challenges for grid stability and reliability. On the other hand, advancements in energy storage technologies and grid management systems are enabling more efficient integration of renewables into the energy mix. By investing in smart grid infrastructure and energy storage solutions, countries can optimize the use of renewable energy and enhance overall system performance.

Community engagement and stakeholder involvement are essential for the successful implementation of renewable energy projects. By involving local communities in the planning and decision-making process, developers can build trust, address concerns, and foster a sense of ownership. Transparent communication, benefit-sharing mechanisms, and capacity-building initiatives can enhance community support and ensure that renewable energy projects contribute to local development goals.

The environmental benefits of renewable energy extend beyond emissions reductions. By reducing air and water pollution, renewable energy can lead to improved public health outcomes and reduced healthcare costs. Cleaner air and water contribute to a healthier population, reducing the economic burden of pollution-related illnesses and enhancing overall quality of life. These benefits underscore the importance of considering both environmental and economic factors in the development of renewable energy projects.

Innovation and technological advancements continue to drive the evolution of renewable energy. Research and development efforts are focused on improving efficiency, reducing costs, and enhancing performance. Advances in energy storage, grid integration, and digital technologies are enabling more efficient and flexible energy systems, maximizing the economic and environmental benefits of renewable energy. By embracing innovation, the renewable energy sector can unlock new opportunities for growth and competitiveness.

The transition to renewable energy presents an opportunity to redefine economic growth in a sustainable and inclusive manner. By prioritizing renewable energy development, countries can create a more resilient and equitable economy that benefits all stakeholders. The balance between environmental and economic considerations is not just about trade-offs; it is about creating a sustainable future that harmonizes economic, environmental, and social goals.

The journey towards a renewable energy future is a shared responsibility that requires collaboration and commitment from governments, industry, communities, and individuals. By working together to address environmental and economic challenges and embrace innovative solutions, we can unlock the full potential of renewable energy and create a cleaner, more prosperous world for future generations.

Real-World Applications and Successes

Renewable energy has moved beyond theoretical discussions and experimental stages to become a vital component of the global energy landscape. Real-world applications and successes demonstrate the transformative power of renewable technologies, showcasing their potential to meet energy demands while addressing environmental and economic challenges. This chapter delves into various examples of how renewable energy is being effectively implemented across the globe, highlighting the diverse ways in which these technologies are making a tangible impact.

One of the most compelling success stories in renewable energy is the widespread adoption of solar power. Countries like Germany, China, and the United States have made significant strides in harnessing solar energy, driven by technological advancements and supportive policies. Germany's Energiewende, or energy transition, has been instrumental in promoting solar energy, resulting in a substantial increase in solar capacity. This transition has not only reduced the country's reliance on fossil fuels but also created jobs and stimulated economic growth. Similarly, China's aggressive investment in solar technology has positioned it as a global leader in solar panel production and deployment, contributing to a significant reduction in solar energy costs worldwide.

Wind energy is another area where real-world applications have demonstrated remarkable success. Denmark, for instance, has become a pioneer in wind energy, with wind turbines generating over 40% of the country's electricity. This achievement is the result of decades of investment in wind technology, favorable government policies, and a strong commitment to renewable energy. The success of Denmark's wind energy sector has inspired other countries to follow suit, with the United Kingdom, the United States, and India making significant investments in wind power infrastructure. Offshore wind farms, in particular, are gaining traction as a viable solution for harnessing wind energy in regions with limited land availability.

Biomass energy has also found success in various real-world applications, particularly in regions with abundant agricultural and forestry resources. Sweden, for example, has effectively utilized biomass energy to reduce its dependence on fossil fuels and achieve a high level of energy self-sufficiency. The country's extensive use of biomass for district heating and electricity generation has contributed to its reputation as a leader in sustainable energy practices. In Brazil, the production of biofuels from sugarcane has become a cornerstone of the country's energy strategy, providing a renewable alternative to gasoline and reducing greenhouse gas emissions.

Hydropower remains one of the most established and widely used forms of renewable energy, with countries like Norway, Canada, and Brazil relying heavily on hydropower for electricity generation. Norway, in particular, generates nearly all of its electricity from hydropower, taking advantage of its abundant water resources and mountainous terrain. This reliance on hydropower has allowed Norway to maintain low electricity prices and achieve a high level of energy security. In Canada, hydropower accounts for a significant portion of the country's electricity supply, supporting both domestic consumption and exports to neighboring countries.

Geothermal energy, while less widespread than other renewable sources, has proven successful in regions with significant geothermal resources. Iceland is a prime example, with geothermal energy providing a substantial portion of the country's electricity and heating needs. The utilization of geothermal energy has enabled Iceland to achieve near-total reliance on renewable energy, reducing its carbon footprint and enhancing energy security. The success of geothermal energy in Iceland has spurred interest in other regions with geothermal potential, such as the United States, the Philippines, and Kenya.

The integration of renewable energy into existing energy systems is another area where real-world applications have demonstrated success. Hybrid energy systems, which combine multiple renewable sources, are becoming increasingly popular as a means of optimizing energy production and enhancing grid stability. For instance, the combination of solar and wind energy in hybrid systems can provide a more consistent and reliable energy supply, addressing the intermittency challenges associated with individual renewable sources. These systems are particularly beneficial in remote or off-grid locations, where access to traditional energy infrastructure is limited.

Community-based renewable energy projects are gaining momentum as a way to empower local communities and promote sustainable development. These projects involve the collective ownership and management of renewable energy systems, allowing communities to generate their own power and reduce reliance on external energy sources. In countries like Germany and Denmark, community wind farms and solar cooperatives have become popular models for local energy production, fostering a sense of ownership and engagement among community members. These projects not only provide clean energy but also create economic opportunities and strengthen social cohesion.

The success of renewable energy in real-world applications is not limited to electricity generation. Renewable technologies are also making significant contributions to the transportation sector, with electric vehicles (EVs) and biofuels offering sustainable alternatives to traditional fossil fuels. The adoption of EVs is accelerating worldwide, driven by advancements in battery technology, government incentives, and growing consumer awareness. Countries like Norway and the Netherlands are leading the charge, with high rates of EV adoption and ambitious targets for phasing out internal combustion engines. Biofuels, such as ethanol and biodiesel, are also playing a crucial role in reducing emissions from the transportation sector, providing a renewable alternative to gasoline and diesel.

The global transition to renewable energy is not without its challenges, but the successes achieved thus far demonstrate the potential for these technologies to transform the energy landscape. By learning from real-world applications and building on existing successes, countries can continue to expand their renewable energy capacity and move towards a more sustainable and resilient energy future.

The journey towards a renewable energy future is a shared responsibility that requires collaboration and commitment from governments, industry, communities, and individuals. By working together to address environmental and economic challenges and embrace innovative solutions, we can unlock the full potential of renewable energy and create a cleaner, more prosperous world for future generations.

Future Directions in Biomass Energy

Biomass energy, a cornerstone of the renewable energy sector, is poised for significant advancements as the world seeks sustainable solutions to meet growing energy demands. The future of biomass energy is shaped by technological innovations, policy developments, and evolving market dynamics. Understanding these future directions is crucial for stakeholders looking to harness the full potential of biomass as a renewable energy source.

One of the most promising future directions in biomass energy is the development of advanced conversion technologies. Traditional methods, such as direct combustion, are being supplemented by more efficient and versatile processes like gasification and pyrolysis. Gasification involves converting biomass into a syngas, which can be used for electricity generation or as a feedstock for producing chemicals and fuels. Pyrolysis, on the other hand, involves heating biomass in the absence of oxygen to produce bio-oil, which can be refined into transportation fuels or used as a chemical feedstock. These advanced technologies offer higher efficiency and lower emissions, making them attractive options for future biomass energy projects.

The integration of biomass energy with other renewable sources is another key trend shaping the future of the sector. Hybrid systems that combine biomass with solar, wind, or hydropower can provide a more stable and reliable energy supply, addressing the intermittency challenges associated with renewable energy. For instance, biomass can be used as a backup power source when solar or wind energy is unavailable, ensuring a continuous energy supply. These integrated systems can optimize energy production, reduce costs, and enhance grid stability, making renewable energy more competitive with traditional fossil fuels.

Sustainability is a critical consideration in the future development of biomass energy. The sourcing of biomass feedstocks must be managed carefully to avoid negative environmental impacts, such as deforestation, soil degradation, and loss of biodiversity. Sustainable biomass production involves practices like reforestation, crop rotation, and the use of agricultural residues, which help maintain ecosystem health and productivity. Certification schemes and sustainability standards play a crucial role in ensuring that biomass energy contributes positively to environmental and social goals.

The role of policy and regulation in shaping the future of biomass energy cannot be overstated. Governments around the world are implementing policies and incentives to promote the development and deployment of biomass energy. These include tax credits, grants, and subsidies that reduce the financial burden on developers and encourage investment in biomass projects. Additionally, renewable energy mandates and carbon pricing mechanisms create a favorable market environment for biomass energy, driving demand and supporting the transition to a low-carbon economy.

Technological innovation continues to drive the evolution of biomass energy. Advances in biotechnology, for example, are enabling the development of genetically modified crops with higher yields and improved resistance to pests and diseases. These crops can provide a more reliable and sustainable source of biomass feedstocks, enhancing the economic viability of biomass energy projects. Furthermore, research into new materials and catalysts is improving the efficiency of biomass conversion processes, reducing costs and emissions.

The potential for biomass energy to contribute to the circular economy is another exciting future direction. By utilizing waste materials, such as agricultural residues, food waste, and municipal solid waste, biomass energy can help close the loop in resource use and reduce the environmental impact of waste disposal. This approach not only provides a sustainable energy source but also supports waste management and resource recovery efforts, contributing to a more sustainable and resilient economy.

The future of biomass energy is also shaped by the growing demand for biofuels in the transportation sector. As countries seek to reduce greenhouse gas emissions and transition to cleaner transportation options, biofuels like ethanol and biodiesel are gaining traction as renewable alternatives to gasoline and diesel. The development of advanced biofuels, such as cellulosic ethanol and algae-based fuels, offers the potential for even greater emissions reductions and sustainability benefits. These fuels can be produced from non-food feedstocks, reducing competition with food production and enhancing the overall sustainability of biofuel production.

Community engagement and stakeholder involvement are essential for the successful implementation of future biomass energy projects. By involving local communities in the planning and decision-making process, developers can build trust, address concerns, and foster a sense of ownership. Transparent communication, benefit-sharing mechanisms, and capacity-building initiatives can enhance community support and ensure that biomass energy projects contribute to local development goals.

The future of biomass energy is bright, with ongoing research and development efforts poised to deliver even greater advancements. As the world transitions to a more sustainable energy system, biomass energy will continue to play a vital role in meeting global energy needs. By embracing innovation and adopting best practices, the biomass energy sector can enhance its contribution to a cleaner, more resilient energy future.

The journey towards sustainable energy is a shared responsibility that requires the collective efforts of governments, industry, communities, and individuals. By working together to address environmental and economic challenges and embrace innovative solutions, we can unlock the full potential of biomass energy and create a cleaner, more prosperous world for future generations.